IAN RIDPATH

LONGMAN ILLUSTRATED DICTIONARY OF ASTRONOMY & ASTRONAUTICS

the terminology of space

LONGMAN  YORK PRESS

YORK PRESS
Immeuble Esseily, Place Riad Solh, Beirut.

LONGMAN GROUP UK LIMITED
Burnt Mill, Harlow, Essex.

© Librairie du Liban 1987

All rights reserved. No part of this publication may be reproduced, stored in a retrieval system, or transmitted in any form or by any means, electronic, mechanical, photocopying, recording, or otherwise, without the prior permission of the copyright owner.

First published 1987

ISBN 0 582 89381 X

Illustrations by Industrial Art Studio
Photocomposed in Britain by Prima Graphics, Camberley, Surrey, England.
Printed and bound in Lebanon by Typopress, Beirut.

Contents

How to use the dictionary	*page* 5
Astronomy	
Celestial sphere Coordinates; ecliptic; great circle; motion; angular measure	8
Orbits Ellipses; Kepler's laws; nodes; motion; types of orbit	16
Solar system Planets; rotation; Bode's law	24
The Sun	28
The Moon	33
Tides and Eclipses Tides; eclipses	37
Earth and its environment General; atmosphere; aurorae; light	43
Time and timekeeping Days; solar time; time zone; atomic time; calendars; years; months; timekeeping	48
Planets and moons	57
Comets	60
Meteors	62
Asteroids	64
Meteorites	66
Stars Brightness; stellar evolution; stellar death; variable stars; double stars; star clusters; astrometry; position, motion & distance; brightest & nearest; charts & catalogues	68
Nebulae	98
Milky Way	101
Galaxies Types of galaxy; classification; clusters of galaxies; radio galaxy	104
The Universe Expansion; evolution	108
Astrophysics Gravity; relativity; Doppler effect; atoms; elements; electromagnetic radiation; the spectrum	112

4 · CONTENTS

Astronomical instruments   123
  Optics; focus; aberration; telescopes; spectroscope;
  photoelectricity; radio astronomy; observatories

History of astronomy   138

Astronautics

  Rockets   141
    Propulsion; rocket types, propellants; engines; launching

  Space flight   152
    General; tracking; manoeuvring; re-entry, landing;
    astronauts; manned flight; space stations; Space Shuttle;
    Spacelab; rockets

  Satellites   178

  Space probes   184

  Astronomy satellites   188

General words in astronomy   192

Appendixes:
One: Planets and moons   197
Two: Constellations   198
Three: Observatories   199
Four: Space Centres   202
Five: Great names in astronomy and astronautics   204
Six: Space abbreviations and acronyms   209
Seven: Symbols used in astronomy   212
Eight: International System of Units (SI)   213

Index   214

How to use the dictionary

This dictionary contains over 1500 words used in astronomy and astronautics. These are arranged in groups under the main headings listed on pp. 3-4. The entries are grouped according to the meaning of the words to help the reader to obtain a broad understanding of the subject.

At the top of each page the subject is shown in bold type and the part of the subject in lighter type. For example, on pp. 94 and 95:

94 · **STARS**/BRIGHTEST & NEAREST

STARS/CHARTS & CATALOGUES · **95**

In the definitions the words used have been limited so far as possible to about 1500 words in common use. These words are those listed in the 'defining vocabulary' in the *New Method English Dictionary* (fifth edition) by M. West and J. G. Endicott (Longman 1976). Words closely related to these words are also used: for example, *characteristic*, defined under *character* in West's *Dictionary*.

In addition to the entries in the text, the dictionary has several useful appendixes which are detailed in the Contents list and are to be found at the back of the dictionary.

1. To find the meaning of a word

Look for the word in the alphabetical index at the end of the book, then turn to the page number listed.

In the index you may find words with a number at the end. These only occur where the same word appears more than once in the dictionary in different contexts. For example, **Mercury**

Mercury^1 is a planet;

Mercury^2 is an American spacecraft.

The description of the word may contain some words with arrows in brackets (parentheses) after them. This shows that the words with arrows are defined near by.

(↑) means that the related word appears above or on the facing page;

(↓) means that the related word appears below or on the facing page.

A word with a page number in brackets after it is defined elsewhere in the dictionary on the page indicated. Looking up the words referred to may help in understanding the meaning of the word that is being defined.

In some cases more than one meaning is given for the same word. Where this is so, the first definition given is the more (or most) common usage of the word. The explanation of each word usually depends on knowing the meaning of a word or words above it. For example, on p. 68 the meaning of *apparent magnitude*, *absolute magnitude*, and the words that follow depends on the meaning of the word *magnitude*, which appears above them. Once the earlier words have been read those that follow become easier to understand. The illustrations have been designed to help the reader understand the definitions but the definitions are not dependent on the illustrations.

2. To find related words

Look in the index for the word you are starting from and turn to the page number shown. Because this dictionary is arranged by ideas, related words will be found in a set on that page or one nearby. The illustrations will also help to show how words relate to one another.

For example, words relating to galaxies are on pp. 104-107. On p. 104 *galaxy* is followed by words used to describe different by types of galaxy and illustrations showing a spiral galaxy and a barred spiral galaxy; p. 105 continues to explain and illustrate galaxies with entries on Hubble classification and tuning fork diagram; p. 106 explains clusters of galaxies and p. 107 gathers together the remaining words on galaxies with words relating to radio galaxies.

3. As an aid to studying or revising

The dictionary can be used for studying or revising a topic, or more simply to refresh your memory. For example, to revise your knowledge of the Sun, you would look up *Sun* in the alphabetical index. Turning to the page indicated, p. 28, you would find *Sun*, *solar constant*, *insolation*, *heliocentric*, *photosphere*, and so on; on p. 29 you would find *granulation*, *sunspot*, *pore*, and so on; p. 30 you would find *Sporer's law*, *butterfly diagram*, etc.

In this way, by starting with one word in a topic you can revise all the words that are important to this topic.

4. To find a word to fit a required meaning

It is almost impossible to find a word to fit a meaning in most dictionaries, but it is easy with this book. For example, if you had forgotten the word for two stars at the same distance from us held together by gravity, all you would have to do would be to look up *double star* in the alphabetical index and turn to the page indicated, p. 85. On reading the definition you would be referred to *physical double* on the next page and there you would find the word with a diagram to illustrate its meaning.

5. Abbreviations used in the definitions

abbr	abbreviation	p.	page
adj	adjective	pl	plural
e.g.	*exempli gratia* (for example)	pp.	pages
etc	*et cetera* (and so on)	sing.	singular
i.e.	*id est* (that is to say)	v	verb
n	noun	=	the same as

N.B. billion = 10^9 (1,000,000,000)

THE
DICTIONARY

8 · CELESTIAL SPHERE/COORDINATES

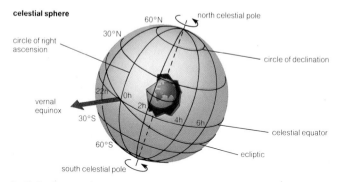

celestial sphere

- **celestial sphere** an imaginary sphere with the Earth at its centre, on which all objects in the sky seem to lie. The celestial sphere appears to turn once each day, but this is actually a result of the Earth spinning on its axis.
- **coordinate** (*n*) a value, such as a distance or angle in a given direction (up, down, left, right) that is used with one or more others to describe the position of an object.
- **position** (*n*) the place at which an object lies. On the celestial sphere, the position of a star, planet or other celestial body is given by its coordinates (↑) in right ascension (↓) and declination (↓).
- **right ascension** a coordinate (↑) on the celestial sphere like longitude on Earth. Right ascension is measured in hours, minutes and seconds eastwards along the celestial equator (↓) starting from zero at the vernal equinox (↑); symbol α.
- **declination** (*n*) a coordinate (↑) on the celestial sphere like latitude on Earth. Declination is measured in degrees north (+) or south (−) of the celestial equator (↓); symbol δ.
- **celestial equator** the celestial equator lies on the celestial sphere directly above the Earth's equator. It thus appears directly overhead to an observer at the Earth's equator. The celestial equator lies exactly half way between the celestial poles (↓).

CELESTIAL SPHERE/ECLIPTIC · 9

celestial pole the celestial poles lie on the celestial sphere directly above the Earth's poles. The celestial poles are the points around which the celestial sphere appears to turn. To an observer standing at the Earth's pole, the celestial pole will lie directly overhead.

pole (*n*) a point on the celestial sphere whose angular distance (p. 15) is 90 degrees from a given great circle (p. 10), e.g. the celestial poles (↑) are 90 degrees north and south of the celestial equator (↑), the ecliptic (↓) poles are 90 degrees north and south of the ecliptic, etc.

ecliptic (*n*) the path that the Sun seems to follow around the celestial sphere each year. As the Earth orbits the Sun, the Sun appears to move along the ecliptic. The ecliptic is a great circle (p. 10) inclined at about 23½ degrees to the celestial equator (↑).

pole of the ecliptic one of two points whose angular distances are 90 degrees north and 90 degrees south of the ecliptic (↑).

obliquity of the ecliptic the angle at which the ecliptic (↑) is tilted to the celestial equator (↑), a result of the tilt of the Earth's axis. This angle is about 23½ degrees, but it changes slightly with time because of the effect of the gravitational (p. 112) pulls of the Sun and Moon.

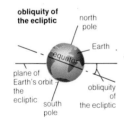

equinox (*n*) one of the two points on the celestial sphere at which the ecliptic (↑) crosses the celestial equator (↑). At an equinox, the Sun lies exactly on the celestial equator so that night and day are equal in length the world over. **equinoctial** (*adj*).

vernal equinox the equinox (↑) that occurs around 21 March each year when the Sun crosses the celestial equator (↑) moving north.

spring equinox = vernal equinox (↑).

autumnal equinox the equinox (↑) that occurs around 23 September each year, when the Sun crosses the celestial equator (↑) moving south.

first point of Aries another name for the vernal equinox (↑), so called because in the past this point lay in the constellation (p. 95) of Aries. Precession (p. 13) has now carried the vernal equinox into the constellation of Pisces.

solstice (*n*) one of two points on the celestial sphere at which the Sun reaches either its greatest declination (p. 8) north (+23½ degrees) or its greatest declination south (−23½ degrees) of the celestial equator (p. 8). **solstitial** (*adj*).

winter solstice the solstice (↑) that occurs around 22 December each year, when the Sun is about 23½ degrees south of the celestial equator (p. 8).

summer solstice the solstice (↑) that occurs around 21 June each year, when the Sun lies about 23½ degrees north of the celestial equator (p. 8).

tropic (*n*) either of the two latitudes on Earth at which the Sun appears farthest north or south during its yearly path along the ecliptic (p. 9).

tropic of Cancer the latitude on Earth, about 23½ degrees north, at which the Sun lies directly overhead at the summer solstice (↑).

tropic of Capricorn the latitude on Earth, about 23½ degrees south, at which the Sun appears directly overhead at the winter solstice (↑).

great circle any circle on the celestial sphere that has the Earth at its centre, e.g. the celestial equator (p. 8), the ecliptic (p. 9), and all lines of right ascension (p. 8). Circles that are not great circles are known as small circles (↓).

small circle any circle on the celestial sphere that does not have the Earth at its centre. Lines of declination (p. 8), apart from the celestial equator (p. 8), are small circles.

zenith (*n*) the point on the celestial sphere that is directly overhead to an observer. **zenithal** (*adj*).

nadir (*n*) the point on the celestial sphere that is directly below an observer. The nadir lies 180 degrees from the zenith (↑).

horizon (*n*) the great circle (↑) on the celestial sphere 90 degrees from the observer's zenith (↑).

meridian (*n*) the great circle (↑) on the celestial sphere that joins the north and south points on the observer's horizon and also passes through the zenith (↑). The meridian passes through the celestial poles (p. 9).

tropic

great and small circles

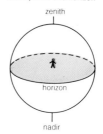
zenith, nadir and horizon

meridian and transit

CELESTIAL SPHERE/GREAT CIRCLE · 11

hour angle, hour circle

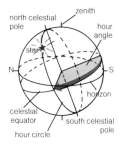

altitude and azimuth

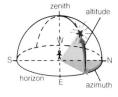

circumpolar

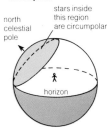

transit[1] (*n*) the moment when a celestial object crosses the meridian (↑).

hour circle a great circle (↑) that passes through an object on the celestial sphere and through the celestial poles (p. 9).

hour angle the angle between the meridian (↑) and the hour circle (↑) that passes through a celestial object. Hour angle is measured westwards on the celestial sphere. The hour angle gives the time since the object last crossed the meridian, e.g. an object on the meridian has zero hour angle; an object that crossed the meridian an hour ago has an hour angle of one hour, etc.

polar distance the angular distance (p. 15) of an object from either the north or south celestial poles (p. 9).

colure (*n*) a great circle (↑) that passes either through the celestial poles (p. 9) and the equinox (p. 9) points (*equinoctial colure*), or through the celestial poles and the solstice (p. 10) points (*solstitial colure*).

altitude[1] (*n*) the angle of a celestial object above the horizon.

azimuth (*n*) the direction of an object, measured in degrees around the observer's horizon clockwise from north.

zenith distance the angular distance (p. 15) of a celestial object from the zenith (↑).

culmination (*n*) the greatest altitude (↑) above the horizon that a star reaches. An object at culmination lies on the meridian (↑).

circumpolar (*adj*) of stars that do not rise or set during the night as seen from a given place on Earth, but instead circle around the celestial pole (p. 9). Circumpolar stars have a polar distance (↑) that is less than the observer's latitude.

rise (*v*) to move above the horizon.

set (*v*) to move below the horizon.

pole star the star nearest to the celestial pole (p. 9) that is visible to the naked eye. The north pole star is Polaris, magnitude (p. 68) 2.2. The southern pole star is Sigma (σ) Octantis, magnitude 5.5. Both pole stars lie about 1 degree from the actual celestial pole.

12 · CELESTIAL SPHERE/COORDINATES

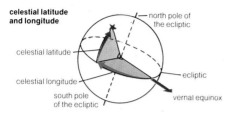

celestial latitude and longitude

celestial latitude a coordinate (p. 8) of an object on the celestial sphere, measured in degrees north or south of the ecliptic (p. 8). Celestial latitude is measured along the great circle (p. 10) passing through the object and the pole of the ecliptic (p. 9); symbol β. *See also* celestial longitude (↓).

celestial longitude a coordinate (p. 8) of an object on the celestial sphere measured along the ecliptic (p. 9) in degrees east of the vernal equinox (p. 9). Celestial latitude (↑) and longitude are sometimes used to give the positions of the Sun, Moon and planets, but are less commonly used than right ascension (p. 8) and declination (p. 8); symbol λ.

geocentric coordinates a set of values which give the position of an object on the celestial sphere as measured from the centre of the Earth. *See also* topocentric coordinates (↓).

topocentric coordinates a set of values which give the position of an object on the celestial sphere as measured from the surface of the Earth. For distant objects such as stars and galaxies there is no noticeable difference between topocentric and geocentric coordinates (↑). But for objects in the solar system the fact that an observer is not at the centre of the Earth makes a difference that must be corrected for in exact measurements.

horizontal parallax the difference between the geocentric coordinates (↑) and the topocentric coordinates (↑) for an object that is on the observer's horizon.

equatorial horizontal parallax the horizontal parallax (↑) of an object seen by an observer on the equator.

diurnal parallax the change in an object's topocentric coordinates (↑) during the day as the observer is carried around by the spin of the Earth.

geocentric parallax = diurnal parallax (↑).

heliocentric coordinates a set of values which give the position of an object as measured from the centre of the Sun. Heliocentric coordinates are often used when describing the positions of bodies in the solar system.

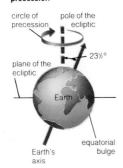

precession

precession (*n*) a movement of the Earth, similar to (but very much slower than) the way in which a spinning top swings around when its axis of rotation is not upright. Because of precession, the position that the Earth's axis points to on the celestial sphere moves around a complete circle every 25,800 years. The circle of precession is centred on the pole of the ecliptic (p. 9). As a result of precession, the position of the celestial poles (p. 9) is continuously changing, as is the position of the equinoxes (p. 9). Precession is caused by the gravitational (p. 112) pulls of the Moon, Sun and the planets on the Earth. Also known as **precession of the equinoxes.** *See also* lunisolar precession (↓), planetary precession (↓). **precess** (*v*).

lunisolar precession the effect of precession (↑) caused by the pulls of the Moon's and Sun's gravity (p. 112) on the Earth. Lunisolar precession is the main cause of the Earth's precession.

planetary precession the effect of precession (↑) caused by the gravitational (p. 112) pulls of the planets on the Earth. Planetary precession has a much smaller effect than lunisolar precession (↑).

Platonic year the time taken for the Earth's poles to complete one circle of precession (↑), i.e. 25,800 years.

nutation (*n*) a slight side-to-side movement of the Earth's axis in space, added to the movement of precession (↑). Nutation causes the Earth's poles to move from side to side by about 9 seconds of arc (p. 15) every 18.6 years.

direct motion (1) the movement of a body such as a planet from west to east on the celestial sphere; (2) the movement of a body in orbit around another from west to east; (3) the spin of a body on its axis from west to east. Direct motion is the normal movement in the solar system, opposite to retrograde motion (↓).
prograde motion = direct motion (↑).

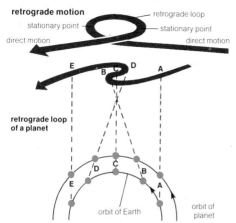

retrograde motion (1) the movement of a body from east to west on the celestial sphere; (2) the movement of a body in its orbit from east to west; (3) the spin of a body on its axis from east to west. Retrograde motion is opposite to direct motion (↑), the normal direction of movement in the solar system. Planets farther from the Sun than the Earth appear to move retrograde at times when the Earth catches them up and overtakes them in its orbit around the Sun. The effect is similar to that when a faster train overtakes a slower one; as seen from the faster train, the slower one seems to be going backwards.

stationary point a position at which an object does not seem to move against the stars when it is changing from direct motion (↑) to retrograde motion (↑) and back again.

CELESTIAL SPHERE / ANGULAR MEASURE · 15

position angle

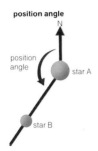

- **position angle** the direction in which one object, e.g. a star, lies on the celestial sphere relative to another object. Position angle is measured in degrees from north via east, south, west and back again to north. Note that east on the celestial sphere is the direction towards the eastern horizon.
- **apparition** (*n*) a length of time during which an object in the solar system, e.g. a planet, is well placed to be seen from Earth in the night sky. The word is not used for objects such as the Sun and stars that are regularly visible.
- **heliacal rising** the occasion when a star or planet can first be seen rising in the morning sky after having been too close to the Sun to be visible. The word is sometimes also used to describe the rising of a celestial object at the same time as the Sun.
- **heliacal setting** the last time that an object can be seen before it gets too close to the Sun to be visible.
- **angular diameter** the apparent size of an object on the celestial sphere, measured in degrees, minutes and seconds of arc (↓).
- **angular distance** the distance between two objects on the celestial sphere, measured in degrees, minutes and seconds of arc (↓).
- **angular measure** angles on the celestial sphere are usually measured in degrees, minutes and seconds of arc (↓). There are sixty minutes of arc in a degree, and sixty seconds of arc in an arc minute. The exception is right ascension (p. 8), which is measured in hours, minutes and seconds. There are 15 degrees in one hour of right ascension.
- **degree** (*n*) one three hundred and sixtieth part of a circle; symbol °.
- **arc minute** one sixtieth of a degree; symbol ′. Not to be confused with minutes of time.
- **arc second** one sixtieth of an arc minute; symbol ″. Not to be confused with seconds of time.
- **arc** (*n*) part of a circle.
- **radian** (*n*) the angle made by an arc of a circle that is equal in length to the radius of the circle; symbol rad.

arc

celestial mechanics the study of the movement of objects that are in orbit, e.g. moons, planets and binary stars (p. 85).

astrodynamics (*n.pl.*) a branch of celestial mechanics (↑) that deals with the movement of artificial satellites (p. 152) and space probes.

orbit (*n*) the path of one body in space around another. The shape of the orbit can be an ellipse (↓), a parabola (p. 19) or a hyperbola (p. 19). **orbit** (*v*), **orbital** (*adj*).

trajectory (*n*) the path of a body in space or through the Earth's atmosphere. The word is usually used to describe the paths of rockets and spacecraft rather than that of planets or moons.

ellipse (*n*) a shape like a flattened circle. The obits of most objects in space are elliptical in shape. *See also* focus (↓), eccentricity (p. 18). **elliptical** (*adj*).

focus[1] (*n*) one of the two points that govern the eccentricity (p. 18) of an elliptical (↑) orbit. The two foci lie at equal distances either side of the centre of the ellipse, along the major axis (p. 18). The object being orbited lies at one of the foci; there is nothing at the other focus. The farther apart the two foci lie, the greater is the eccentricity of the ellipse. **foci** (*pl*).

primary[1] (*n*) a larger body that a smaller body orbits around, e.g. the Sun is the Earth's primary. The smaller body is known as the secondary (↓).

secondary[1] (*n*) the smaller body that orbits around a larger one, e.g. the Moon is the Earth's secondary. *See also* primary (↑).

ellipse

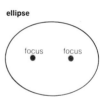

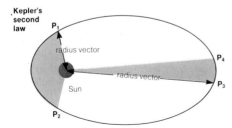

Kepler's second law

Kepler's laws three laws concerning the movement of planets in their orbits around the Sun, discovered by Johannes Kepler. Kepler's first law states that all planets travel around the Sun in ellipses (↑) with the Sun at one focus (↑). The second law states that the radius vector (↓) joining the Sun and the planet moves through equal areas of space in equal lengths of time, meaning that the planet moves fastest in its orbit when it is at perihelion (↓) and slowest when it is at aphelion (p. 18). *On the diagram opposite on p. 16*, the time taken for a planet to go from P_1 to P_2 is the same as from P_3 to P_4. The grey coloured area swept out is the same each time. Kepler's third law states that for each planet in the solar system, the square of its orbital period (p. 20) in years equals the cube of its semimajor axis (p. 18) in astronomical units (p. 24). Put mathematically, the third law becomes $P^2 = a^3$ where P is the orbital period and a is the semimajor axis. It means that the greater the average distance of a planet from the Sun, the longer it takes to complete one orbit. From the third law a planet's distance from the Sun can be calculated once its period is known.

radius vector an imaginary straight line that can be drawn between an object in orbit, e.g. a planet, and the object it is in orbit around, e.g. the Sun.

peri- (*prefix*) relating to the point in a body's orbit that lies closest to the body it is in orbit around.

perigee (*n*) the point at which a body in orbit around the Earth comes closest to the Earth.

perihelion (*n*) the point at which a body in orbit around the Sun comes closest to the Sun; symbol *q*. **perihelia** (*pl*).

periastron (*n*) the point at which an object in orbit around a star comes closest to that star.

perigee and apogee

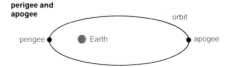

ap-, apo- (*prefixes*) relating to the point in an object's orbit that lies farthest from the body it is in orbit around.

apogee (*n*) the point at which a body in orbit around the Earth is farthest from Earth. *See diagram* (p. 17).

aphelion (*n*) the point at which a body orbiting the Sun is farthest from the Sun; symbol Q. **aphelia** (*pl*).

apastron (*n*) the point at which an object in orbit around a star is farthest from the star.

apsis (*n*) the point in an orbit at which an object is closest to or farthest from the body it is in orbit around. The closest point is known as *periapsis*, and the farthest point is *apoapsis*. In the case of a planet orbiting the Sun, the two apsides are known as perihelion (p. 17) and aphelion (p. 17). For an object orbiting the Earth, the apsides are perigee (p. 17) and apogee (↑). Also known as **apse**. **apsides** (*pl*), **apsidal** (*adj*).

line of apsides the straight line joining the apsides (↑) of an orbit. The line of apsides is the major axis (↓) of the orbit.

apsidal motion the movement of the line of apsides (↑) of an orbit. Apsidal motion is caused by the gravitational (p. 112) pull of another body or, in the case of a binary star (p. 85), if one or both stars are not spherical in shape.

eccentricity (*n*) a measure of the flattening of an ellipse (p. 16), or how much it departs from a circle; symbol e. Ellipses have eccentricities between 0 and 1. When the eccentricity is 0, the shape is a circle; when the eccentricity is 1, the shape is a parabola (↓). An orbit with an eccentricity greater than 1 is a hyperbola (↓). The eccentricity of an ellipse is calculated by dividing the distance between the two foci (p. 16) of the ellipse by the length of the major axis (↓). **eccentric** (*adj*).

major axis the longest diameter of an ellipse (p. 16); one which passes through two foci (p.16).

semimajor axis half the longest diameter of an ellipse (p. 16); symbol a. It is also the average distance of an orbiting object from its primary (p. 16).

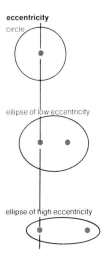

eccentricity
circle

ellipse of low eccentricity

ellipse of high eccentricity

major and minor axis

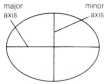

major axis   minor axis

ORBITS/ELLIPSES · 19

parabola

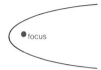

hyperbola

inclination

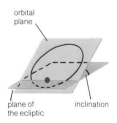

plane of the ecliptic · inclination

minor axis the shortest diameter of an ellipse (p. 16), at right angles to the major axis (↑).

parabola (*n*) a curve that never quite closes in on itself. The sides of a parabola become parallel towards infinity. A parabola can be thought of as an ellipse (p. 16) in which the two foci (p. 16) are infinitely far apart. The orbits of many comets are so highly elliptical that they are difficult to tell from parabolas. **parabolic** (*adj*).

parabolic velocity = escape velocity (↓).

escape velocity the speed at which an object will just break free from the gravitational (p. 112) pull of another body, moving away never to return. An object moving at exactly escape velocity follows a parabolic (↑) trajectory (p. 16). An object whose speed is greater than escape velocity will move on a hyperbolic (↓) trajectory.

hyperbola (*n*) a curve whose sides never close in on themselves but continue always to move apart. The orbit of one object that flies past another without being captured (↓) by it is a hyperbola. **hyperbolic** (*adj*).

hyperbolic velocity the speed of a body that is moving faster than escape velocity (↑).

captured (*adj*) of an object that has become trapped by gravity (p. 112) into an orbit around another body.

elements of an orbit six quantities that describe the orbit of a body in space. They are: (1) the inclination (↓) of the orbit; (2) the longitude of the ascending node (p. 20); (3) either the argument of perihelion (p. 20) or the longitude of perihelion (p. 20); (4) the semimajor axis (↑); (5) the eccentricity (↑); (6) the time of perihelion passage (p. 20). In addition, the period (p. 20) of the orbit may also be given.

inclination (*n*) (1) the angle at which the orbit of an object in the solar system is tilted with respect to the plane of the Earth's orbit, the ecliptic (p. 9); symbol *i*. For an object orbiting the Earth or another planet, the inclination is usually given relative to the equator of the Earth or planet; (2) the angle at which a planet's axis is tilted with respect to the upright. **incline** (*v*).

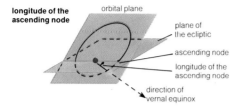

longitude of the ascending node the angle measured eastwards along the ecliptic (p. 9) between the vernal equinox (p. 9) and the ascending node (↓) of an orbit; symbol Ω. The longitude of the ascending node defines one of the two points at which an object's orbit crosses the plane of the Earth's orbit.

node (*n*) one of two points at which the orbit of a body crosses a given plane, usually the plane of the Earth's orbit or the plane of the Earth's equator. At the *ascending node*, the body is moving from south to north; at the *descending node*, it is moving from north to south. **nodical** (*adj*).

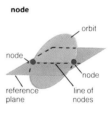

line of nodes the straight line joining the ascending and descending nodes (↑) of an orbit.

regression of nodes the westward movement of the nodes (↑) of an orbit caused by the gravitational (p. 112) pull of other bodies, notably the Sun. The nodes of the Moon's orbit regress once around the ecliptic (p. 9) in 18.6 years.

argument of perihelion the angle between the ascending node (↑) and the perihelion (p. 17) of an orbit, measured in the plane of the orbit and in the direction that the object moves along its orbit; symbol ω. The argument of perihelion defines the direction of the major axis (p. 18) of an orbit.

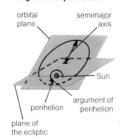

longitude of perihelion the longitude of the ascending node (↑) added to the argument of perihelion (p. 17); symbol $\overline{\omega}$.

time of perihelion passage the date and time at which an object in orbit around the Sun reaches its closest point to the Sun; symbol T.

period[1] (*n*) the length of time taken by a body to go once around its orbit; symbol P.

anomaly

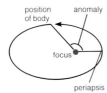

mean motion the average movement of an object along its orbit, usually given in degrees per day; symbol n.

epoch (*n*) a set date for which positions of stars or the elements of an orbit (p. 19) are given. Star positions change over time because of the effect of precession (p. 13). The epoch currently used for star catalogues and atlases is the start of the year 2000.

anomaly (*n*) the angle describing how far a body has moved around its orbit. **anomalistic** (*adj*).

true anomaly the angle between the direction of periapsis (p. 18) and the actual position of an object in its orbit, measured in the direction of the object's movement.

mean anomaly in an orbit, the angle from the direction of periapsis (p. 18) to the position that an object would be at if it moved around its orbit at constant speed, instead of moving faster at periapsis and slower at apoapsis as actually happens.

ephemeris (*n*) a table of predicted positions for a celestial body. **ephemerides** (*pl*).

osculating elements the elements of the orbit (p. 19) that an object in the solar system would follow if the perturbations (↓) caused by all other bodies in the solar system were removed at a given time. Such an *osculating orbit* would be a perfect ellipse (p. 16), but this ideal case is never met with in practice. The paths of all objects in the solar system are not quite perfect ellipses because of the slight gravitational (p. 112) pulls of the other planets. Osculating elements change with time because the planets are moving and so the perturbations they cause are always changing. **osculatory** (*adj*).

perturbation (*n*) a slight effect on the movement of an object caused by the gravitational (p. 112) pull of other bodies. The effect of perturbations means that an object is pulled slightly out of its expected position, so that it does not follow a smooth course. **perturb** (*v*).

residual (*n*) a small difference between the observed and predicted positions of an object in its orbit.

Lagrangian points five places at which a small body can exist in the orbital plane of two much larger bodies. In the case of a planet orbiting the Sun one Lagrangian point called L_1 lies between the planet and the Sun. Point L_2 lies farther from the Sun than the planet, while L_3 lies on the opposite side of the orbit from the planet. In practice, the perturbations (p. 21) caused by the planets will soon pull any objects away from these three points. Of greatest interest are the points L_4 and L_5 that lie 60 degrees ahead of and behind the planet in its orbit. There, objects can remain for ever despite the perturbations of the planets. It is at these two Lagrangian points along the orbit of Jupiter that the Trojan (p. 65) asteroids lie.

libration points = Lagrangian points (↑).

commensurable (*adj*) of the orbital period (p. 20) of one object, e.g. a moon or planet, that is an exact fraction (e.g. one half, one third, etc) of the period of another object that is in orbit around the same body. **commensurability** (*n*).

Lagrangian points

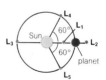

barycentre

barycentre – equal masses

barycentre – unequal masses

barycentre (*n*) strictly speaking, one object does not orbit another; rather, the two bodies orbit their common centre of mass (↓) known as the barycentre. If the two bodies are of the same mass, the barycentre lies exactly half way between them. The larger one body is, the closer to it the barycentre lies. In the case of the Earth–Moon system, the barycentre lies 4,670 km from the centre of the Earth (i.e. about 1,600 km under the surface of the Earth).

centre of mass the point that acts as though all the mass of an object or system of objects were concentrated at it.

synchronous orbit an orbit in which an artificial satellite (p. 152) or a natural moon goes once around a planet in the same time as the planet takes to spin on its own axis.

geosynchronous orbit a synchronous orbit (↑) around the Earth.

ORBITS/TYPES OF ORBIT · 23

geostationary orbit

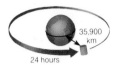

polar orbit

Hohmann orbit
from Earth to Mars

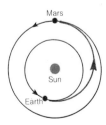

geostationary orbit a synchronous orbit (↑) 35,900 km above the Earth's equator. At that distance, objects orbit once around the Earth in 24 hours, so that they move at the same rate as the Earth spins. Therefore they appear fixed at a point above the Earth's equator.

Clarke orbit = geostationary orbit (↑).

Sun-synchronous orbit an orbit around the Earth or other planet in which a satellite passes over the same part of the planet at the same time each day.

polar orbit an orbit that passes over the poles of a planet or nearly so, i.e. the inclination (p. 19) of the orbit is close to 90 degrees. A satellite in polar orbit passes over every point on the planet below as the planet rotates on its axis.

parking orbit an orbit into which a spacecraft is placed before being sent on to a new trajectory (p. 16), e.g. space probes are placed into parking orbit around the Earth before being sent to the Moon or planets. On arrival at the Moon or a planet, the spacecraft may enter a parking orbit before it, or a part of it, descends to the surface or returns to the Earth.

direct ascent a trajectory (p. 16) in which a spacecraft travels directly from one body to another without first entering a parking orbit (↑).

transfer orbit the trajectory (p. 16) of a spacecraft that is moving between one orbit and another. A special case of a transfer orbit is the Hohmann orbit (↓). Also known as **transfer ellipse.**

Hohmann orbit a trajectory (p. 16) along which a spacecraft needs the least amount of energy to move between one orbit to another one, such as from the Earth to another planet or from low orbit around the Earth to geostationary orbit (↑). A Hohmann orbit just touches the two orbits, without crossing them. A Hohmann orbit saves fuel, but it is the slowest way to go. It is named after Dr Walter Hohmann of Germany who calculated such orbits in 1925.

minimum-energy orbit = Hohmann orbit (↑).

flyby (n) the passage of a spacecraft past a moon or a planet without going into orbit around it.

solar system the Sun and its family of planets and their moons, asteroids and comets. All objects in the solar system are held by the Sun's gravitational (p. 112) pull, which can hold objects out to a distance of about two light years (p. 91), half way to the nearest star.

planet (*n*) a body of rock, metal or gas that does not give out its own light. A planet can have a mass up to about 60 times that of Jupiter before it becomes a star. **planetary** (*adj*).

inferior planet a planet that orbits closer to the Sun than does Earth, i.e. Mercury and Venus.

superior planet a planet whose distance from the Sun is greater than that of the Earth.

terrestrial planet any small, rocky planet. In the solar system, the terrestrial planets are Mercury, Venus, Earth and Mars.

giant planet any planet much larger than the Earth, made of gas. Jupiter, Saturn, Uranus and Neptune are the giant planets of our solar system.

astronomical unit the average distance of the Earth from the Sun, i.e. 149,600,000 km. The astronomical unit is the basic unit of distance in the solar system; symbol AU.

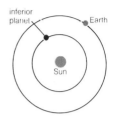

inferior planet

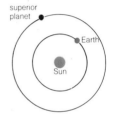

superior planet

astronomical unit

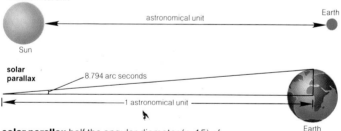

solar parallax half the angular diameter (p. 15) of the Earth's equator as seen at a distance of one astronomical unit (1), i.e. 8.794 seconds of arc (p. 15).

satellite[1] (*n*) any small body that orbits a larger one. All the planets except Mercury and Venus are known to have at least one natural satellite. each.

greatest elongation

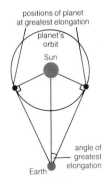

superior and inferior conjunction

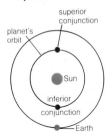

conjunction and opposition

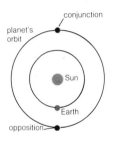

moon = satellite (↑). The Earth's natural satellite, the Moon, is given a capital letter.

sidereal period the time taken for a planet or satellite (↑) to go once around its orbit with respect to the stars.

synodic period the time taken for a planet or satellite (↑) to return to the same position in the sky as seen from Earth, e.g. the time from one opposition (↓) to the next. The synodic period differs from the sidereal period (↑) because the Earth itself is moving in orbit around the Sun.

aspect (n) the position of a planet relative to the Sun, as seen from Earth. Major aspects are conjunction (↓), opposition (↓) and, for Mercury and Venus, greatest elongation (↓).

elongation (n) the angle between the Sun and a planet, or between a planet and one of its satellites (↑), as seen from the Earth.

greatest elongation the points at which Mercury and Venus reach their widest angular distance (p. 15) from the Sun.

conjunction (n) the moment when two bodies in the solar system have the same celestial longitude (p. 12) as seen from Earth. Two planets can be in conjunction with each other, or the Moon can be in conjunction with a planet. A superior planet (↑) is said to be in conjunction when it lies directly behind the Sun.

superior conjunction the moment when an inferior planet (↑) lies directly behind the Sun as seen from Earth.

inferior conjunction the moment when an inferior planet (↑) lies directly between the Earth and the Sun.

opposition (n) the moment when a superior planet (↑) lies directly opposite to the Sun in the sky, as seen from Earth. Inferior planets (↑) cannot come to opposition.

syzygy (n) the moment when a planet or the Moon lies directly in line with the Earth and the Sun. In the case of the Moon, syzygy is the time of full Moon (p. 33) and new Moon (p. 33); in the case of a planet, it is the time of conjunction (↑) and opposition (↑).

revolution[1] (n) the movement of one body in orbit around another, e.g. the Earth makes one revolution around the Sun in a year. **revolve** (v).

rotation (n) the spin of a body on its axis. **rotate** (v).

captured rotation the movement of a body that spins on its axis in the same time as it takes to complete one orbital revolution. A body with a captured rotation therefore keeps one face turned all the time towards the object it is in orbit around. Our own Moon and many moons of other planets have a captured rotation.

synchronous rotation = captured rotation (↑).

differential rotation the movement of a body in which different parts spin at different speeds. Bodies made of gas, e.g. the Sun and the giant planets (p. 24), spin faster at the equator than at the poles.

oblateness (n) the flattening of a planet or star, caused by its speed of rotation. The amount of oblateness is calculated by subtracting the polar diameter of the object from its equatorial diameter, and then dividing by the equatorial diameter. **oblate** (adj).

equatorial bulge a result of the rotation of a star or planet, which causes the diameter across the object's equator to be larger than that across its poles. See also oblateness (↑).

polar flattening = oblateness (↑).

disc (n) the face of a planet, satellite (p. 24), or the Sun as seen from Earth.

limb (n) the edge of the disc (↑) of a planet, satellite (↑), or the Sun.

invariable plane a fixed plane in the solar system, given by adding up the angular momentum (↓) of all the planets and their moons. The angular momentum results from both their revolution in their orbits and their rotation on their axes. The invariable plane lies at an angle of just over 1½ degrees to the plane of the ecliptic (p. 9).

angular momentum a measure of the spin of a body. The angular momentum of a body takes into account its mass; the distance of the body from the centre of revolution or the distance of its various parts from the axis of rotation; and its speed of revolution or rotation.

revolution

rotation

captured rotation

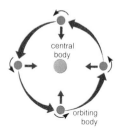

oblateness

albedo (*n*) a measure of the amount of light that is reflected by a planet or satellite (p. 24). If all the incoming sunlight were reflected, the body would have an albedo of 1. If none of the incoming light were reflected, the body would have an albedo of 0 and would appear totally black. The albedo of an object tells how bright its surface is.

Bode's law a simple system of measurement, used by the German astronomer Johann Bode in the 18th century, that describes the distances of the planets from the Sun. Bode's law works as follows. Take the number 0, 3, 6, 12 etc., doubling the number at each step. Add 4 to each number and divide by 10. The result gives the rough distance of each planet from the Sun in astronomical units (p. 24) out as far as Uranus (*see* table). The 'law' breaks down for Neptune and Pluto.

Bode's law

planet	distance given by Bode's law	actual distance from the Sun
	in astronomical units	
Mercury	0.4	0.39
Venus	0.7	0.72
Earth	1.0	1.0
Mars	1.6	1.5
asteroids	2.8	2.8
Jupiter	5.2	5.2
Saturn	10.0	9.5
Uranus	19.6	19.2
Neptune	38.8	30.1
Pluto	77.2	39.5

nebular hypothesis a theory formed at the end of the 18th century by Laplace to explain the origin of the solar system. On this theory, the Sun and planets were born from a giant cloud of gas and dust in space, known as the *solar nebula*.

accretion (*n*) the growth of a body by the addition of matter from outside. The planets are thought to have been built up by accretion from a cloud of gas and dust around the Sun. **accrete** (*v*).

protoplanet (*n*) an object that is growing into a planet by accretion (↑).

planetesimal (*n*) any small body that comes together with others to build up a planet by accretion (↑).

28 · THE SUN

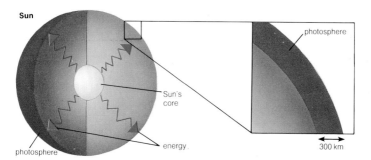

Sun (*n*) the star that lies at the centre of the solar system. The Sun is a glowing ball of gas 1,400,000 km in diameter. The temperature of its surface, the photosphere (↓), is 5,800 K, but it is much hotter at the centre, about 15 million kelvins. The Sun consists of about two-thirds hydrogen, one-third helium. All other elements (p. 116) make up only about 1 per cent of the Sun. Its mass is 2×10^{30} kg, its volume is 1.3 million times that of the Earth, and its density is 1.4 times that of water. Other stars are sometimes also called suns.

solar (*adj*) of the Sun.

solar constant the amount of energy from the Sun that falls on a given area at the edge of the Earth's atmosphere, i.e. about 1.35 kilowatts per square metre.

insolation (*n*) the amount of solar energy that falls on a surface.

heli-, helio- (*prefixes*) referring to the Sun.

heliocentric (*adj*) with the Sun at the centre, e.g. the planets are in heliocentric orbit.

photosphere (*n*) the surface of the Sun. The photosphere is not solid, but is a layer of gas about 300 km thick that gives out light and heat. **photospheric** (*adj*).

limb darkening an effect in which the edge of the Sun appears fainter than the centre of the disc (p. 26). The effect occurs because the light we see at the Sun's limb (p. 26) comes from higher in the Sun's photosphere (↑) where the temperature is less.

limb darkening

THE SUN · 29

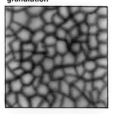
granulation

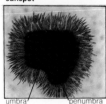
sunspot / umbra / penumbra

granulation (*n*) the grainy appearance of the photosphere (↑), caused by pockets of hot gas rising and falling like water boiling in a pot. Each *granule* is about 1,000 km across. Also known as **rice-grain effect**. **granule** (*n*).

sunspot (*n*) a dark marking on the surface of the Sun. The gases in a sunspot are up to 2,000 kelvins cooler than the rest of the photosphere (↑). Sunspots can be 100,000 km or more in diameter, and often appear in groups. A sunspot lasts for a few weeks before dying out. Sunspots are believed to be caused by lines of magnetic force on the Sun that stop heat from reaching the surface.

pore (*n*) a small sunspot (↑) not much bigger than a granule (↑), about 1,000 km across.

umbra^2 (*n*) the darkest, central part of a sunspot (↑).

penumbra^2 (*n*) the lighter-coloured outer part of a sunspot (↑).

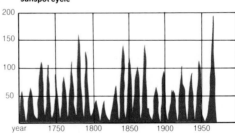
sunspot cycle

sunspot cycle the rise and fall in the number of sunspots (↑) seen at any one time. At best, 100 or more spots can be seen on the Sun, but at the least active part of the cycle no spots are seen for many days. The sunspot cycle can last from 8 to 16 years, but its average length is 11 years. Other events on the Sun also follow the same cycle.

sunspot maximum the time in the sunspot cycle (↑) when most sunspots (↑) are seen.

sunspot minimum the time in the sunspot cycle (↑) when fewest sunspots (↑) are seen.

Spörer's law the fact that the first sunspots
(p. 29) in each new sunspot cycle (p. 29) appear
at latitudes 30 degrees to 40 degrees on the
Sun, while later spots in the cycle appear
closer and closer to the Sun's equator. The last
sunspots of each cycle appear a few degrees
either side of the Sun's equator.

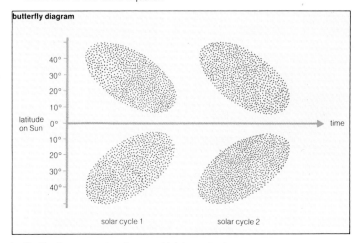

butterfly diagram a drawing on which is marked
the latitude of sunspots (p. 29) throughout each
sunspot cycle (p. 29), showing how they get
closer to the equator during the cycle, as
stated by Spörer's law (↑). The diagram gets its
name because its shape looks like the wings of
a butterfly.

Wilson effect when a sunspot (p. 29) is seen
near the edge of the Sun the central umbra
(p. 29) seems to lie lower than the penumbra
(p. 29) around it, as though the whole spot were
dish-shaped. This is the Wilson effect.

facula (n) a bright patch on the photosphere
(p. 28), consisting of hotter gas than the
photosphere around it. Faculae appear in areas
where sunspots (p. 29) later form, and they live
on for some weeks after the sunspots have
died out. **faculae** (pl).

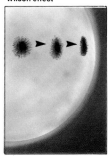

Wilson effect

chromosphere (*n*) a layer of gas about 10,000 km deep that lies above the Sun's photosphere (p. 28). The gases of the chromosphere are much thinner than those of the photosphere and thus give out much less light; we can see straight through the chromosphere to the photosphere. The chromosphere is visible at an eclipse of the Sun when the light from the photosphere is blocked by the Moon.

flash spectrum the spectrum (p. 119) of light from the chromosphere (↑), so named because the chromosphere flashes out for only a few seconds at the beginning and end of a total eclipse (p. 40).

spicule (*n*) a tongue of gas that shoots upwards from the chromosphere (↑) to 10,000 km or so. Each spicule lasts for only a few minutes.

plage (*n*) an area of brighter, hotter gas in the chromosphere (↑), similar to faculae (↑). Plages appear in the same areas as sunspots (p. 29).

flocculus (*n*) any of several, different types of marking in the chromosphere (↑), including plages (↑), filaments (p. 32) and spicules (↑). **flocculi** (*pl*).

supergranulation (*n*) an effect caused by the flow of hot gases in the photosphere (p. 28) and chromosphere (↑). Each *supergranule* is about 30,000 km across, much larger than an ordinary granule (p. 29). Around the edges of supergranules are spicules (↑).

flare[1] (*n*) a sudden burst of light in the chromosphere (↑). Flares brighten within a few minutes and then fade away over an hour or so. Flares happen near sunspots (p. 28). They throw out streams of atomic (p. 116) particles into space.

prominence (*n*) a bright cloud of gas reaching up from the chromosphere (↑) into the corona (p. 32). Prominences are visible around the Sun's edge at total eclipses (p. 40). There are several types of prominence (*see below*).

quiescent prominence a prominence (↑) that lasts for weeks or months, often without changing very much. Quiescent prominences look like arches. They are 100,000 km or more long and over 10,000 km high.

prominences

active prominence a short-lived prominence (p. 31) caused by a flare (p. 31). When a flare occurs, gas is thrown upwards in a *spray prominence*. After the flare has died away, gas from the corona (↓) falls back to the Sun's surface in a *loop prominence*.

filament (*n*) a prominence (p. 31) seen against the face of the Sun, when it appears as a long, dark marking.

corona (*n*) a ring of faint gas around the Sun that starts at the top of the chromosphere (p. 31) and thins out into space. The corona can be seen at a total eclipse (p. 40). Its shape changes with the sunspot cycle (p. 29), being more rounded at sunspot maximum (p. 29) than at sunspot minimum (p. 29). The gas of the corona is at about 2 million K. Gas from the corona flows away from the Sun to form the solar wind (↓).

K corona the brightest part of the Sun's corona (↑), nearest to the Sun. It consists of electrons (p. 116) that scatter light from the photosphere (p. 28). The name K corona comes from the German word *kontinuum* (a sequence).

F corona the outer part of the Sun's corona (↑), fainter than the K corona (↑). It shines by sunlight scattered by particles of dust. The F stands for Fraunhofer (the German physicist).

coronal hole a part of the Sun's corona (↑) in which the gas is much thinner than in the rest.

solar wind a stream of atomic (p. 116) particles flowing outwards from the Sun in all directions at speeds of 1 million $km\,h^{-1}$ and above. The solar wind consists mostly of protons (p. 116) and electrons (p. 116) boiled off from the Sun's corona (↑), plus gas thrown off from the Sun by activity such as flares (p. 31). The solar wind flows through the solar system, past the Earth and the other planets.

heliosphere (*n*) the volume of space through which the solar wind (↑) flows. Its edge is thought to be about 100 astronomical units (p. 24) from the Sun, where the flow of the solar wind is stopped by the gas between the stars.

heliopause (*n*) the outer edge of the heliosphere (↑). It is thought to lie about 100 AU from the Sun.

THE MOON · 33

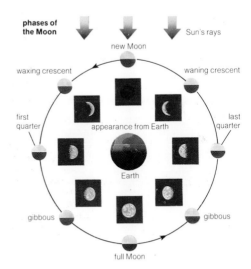

phases of the Moon

- **Moon** (*n*) the Earth's natural satellite (p. 24). The Moon is a rocky body 3,476 km in diameter. Its mass is 81 times less than that of the Earth. The Moon orbits the Earth every 27.32 days at an average distance of 384,400 km. It also spins on its axis in 27.32 days, so that the same side of it faces the Earth all the time. The Moon has no air or water.
- **lunar** (*adj*) of the Moon.
- **cynthion** (*adj*) of the Moon.
- **phase** (*n*) the amount of the sunlit side of the Moon (or any other celestial body) that is visible from Earth. The Moon goes through a series of phases that repeat themselves every 29.53 days, a length of time known as a synodic month (p. 55).
- **new Moon** the time when none of the sunlit side of the Moon is visible from Earth. At new Moon, the Moon and Sun lie in the same direction in the sky.
- **full Moon** the time when all the sunlit side of the Moon is visible from Earth. At full Moon, the Moon lies opposite the Sun in the sky.

34 · THE MOON

lunation (*n*) the time taken for the Moon to go through one set of phases (p. 33), such as from one new Moon (p. 33) to the next. It lasts 29.53 days, and is the same as a synodic month (p. 55).

waxing (*adj*) of the Moon when its phase (p. 33) is increasing from new to full. **wax** (*v*).

waning (*adj*) of the Moon when its phase (p. 33) is decreasing from full to new. **wane** (*v*).

crescent (*n*) the phase (p. 33) of the Moon or a planet between new and first quarter (↓) (*waxing crescent*), or between last quarter (↓) and new (*waning crescent*).

cusp (*n*) one of the points of the crescent (↑) Moon, or of a planet in crescent phase (p. 33).

first quarter the time between new Moon (p. 33) and full Moon (p. 33) when exactly half the sunlit side of the Moon is visible, i.e. the half phase (p. 33) of the waxing (↑) Moon.

last quarter the time between full Moon (p. 33) and new Moon (p. 33) when exactly half the sunlit side of the Moon is visible, i.e. the half phase (p. 33) of the waning (↑) Moon.

third quarter = last quarter (↑).

gibbous (*n*) the phase (p. 33) of the Moon or a planet when it is between half and fully lit.

terminator (*n*) the line dividing the sunlit and dark part of the Moon or a planet. The terminator is the line of sunrise or sunset, separating day from night.

earthshine (*n*) sunlight reflected from the Earth onto the Moon, which makes even the dark part of the Moon faintly visible when at crescent (↑) phase (↑).

crater (*n*) a circular hollow in the surface of the Moon or other body. Craters on the Moon range in size up to 300 km in diameter. Most craters on the Moon and bodies in the solar system were caused by meteorites hitting them, though some are due to volcanoes.

impact crater a crater (↑) formed by a meteorite hitting the ground.

central peak a hill or mountain on the floor of a crater (↑).

terracing (*n*) a step-like effect on the walls of a crater (↑), caused by landslides.

terminator

limb   terminator

crater

central peak   terracing

THE MOON · 35

ray crater

rays

ray (n) a bright streak of ejecta (↓) around young impact craters (↑).
ray crater a crater (↑) with rays (↑).
ejecta (n.pl.) dust and rocks thrown out from an impact crater (↑) or a volcano.
mare (n) a smooth lowland area on the Moon, darker in colour than the much cratered (↑) highlands. The lunar maria were caused by flows of lava in the distant past. **maria** (pl).
terrae (n.pl.) the highlands of the Moon which are lighter in colour than the maria (↑) and much cratered (↑).

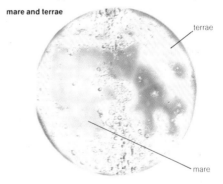
mare and terrae

terrae

mare

mascon (n) an area of denser rock under the surface of the Moon's maria (↑), which makes the Moon's gravitational (p. 112) pull in those areas slightly stronger than normal. The name is short for *mass concentration*.
regolith (n) soil on the surface of the Moon, made from rock that has been broken up by meteorites hitting it.
dome[1] (n) a swelling on the surface of a lunar mare (↑), probably caused by volcanic lava breaking through from underneath.
wrinkle ridge a long, low hill on the surface of a mare (↑), looking like a frozen wave. Wrinkle ridges are thought to have been formed by flows of lava. They can measure several hundred kilometres long, but are only a few hundred metres high.

wrinkle ridge

36 · THE MOON

rille (*n*) a long valley with a flat floor, caused by faults in the Moon's crust.
rima (*n*) = rille (↑). **rimae** (*pl*)
sinuous rille a valley on a lunar mare (p. 35) with many S-shaped bends, like a dried-up river bed. Sinuous rilles are thought to have been formed by flows of lava not running water.
selenology (*n*) the study of the Moon and its surface, i.e. lunar geology.
libration (*n*) an effect that allows us to see slightly more than half of the Moon's surface. Libration in longitude (↓) added to libration in latitude (↓) means that we can see a total of 59 per cent of the Moon's surface.
libration in longitude an effect in which the Moon appears to swing from side to side on its axis. This occurs because the Moon's speed along its orbit varies with its distance from Earth, in accordance with Kepler's laws (p. 17). This swing allows us to see a little way around the east and west edges of the Moon.

sinuous rille

libration in longitude

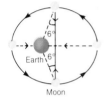

libration in latitude

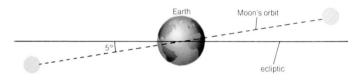

libration in latitude an effect caused by the fact that the Moon's orbit is inclined (p. 19) to the ecliptic (p. 9), so that we can see over the north and south polar regions.
equation of the centre the difference between the true anomaly (p. 21) and the mean anomaly (p. 21) of the Moon in its orbit around the Earth.
evection (*n*) a change in the eccentricity (p. 18) of the Moon's orbit caused by the gravitational (p. 112) pull of the Sun.
variation (*n*) a perturbation (p. 21) of the Moon's motion caused by the gravitational (p. 112) pulls of the Sun and the Earth.

tide (*n*) the rise and fall of the waters of the Earth caused by the gravitational (p. 112) pull of the Moon and the Sun. The Moon has over twice the effect on the tides that the Sun does, because it is much closer to the Earth. **tidal** (*adj*).

high tide the time when the water level of the sea is highest. There are usually two high tides at a given place each day. Between them occur low tides.

low tide the time each day when the water level of the sea is lowest. There are usually two low tides each day.

tidal bulge

tidal bulge the two heaps of water in the seas of Earth caused by the gravitational (p. 112) pulls of the Sun and Moon. As the Earth rotates each day, these two tidal bulges sweep over the Earth, producing high and low tides.

spring tide a very high tide that is produced when the Moon and Sun are pulling in line, so that their effects add up. There are two spring tides each month, at new Moon (p. 33) and at full Moon (p. 33).

spring tides

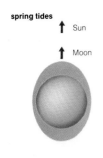

neap tide a tide that has a small range between high and low. Neap tides are produced when the Moon and Sun are pulling on the Earth at right angles, as happens when the Moon is at first quarter (p. 34) and at last quarter (p. 34).

neap tides

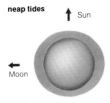

tidal distortion the condition when one body is pulled out of shape by the gravity (p. 112) of another. A small moon can be tidally distorted by the gravitational pull of its parent planet, or two stars close together can tidally distort each other.

tidal friction a force caused by the tides that is slowing down the rotation of the Earth on its axis. Because of tidal friction, the length of the day is increasing by 1.6 seconds every 100,000 years.

38 · TIDES AND ECLIPSES/ECLIPSES

eclipse (n) an event in which light from one body is blocked off by another body. Eclipses of the Sun and Moon can happen only when the Sun and Moon line up with the Earth. This happens at new Moon (p. 33) and full Moon (p. 33). But eclipses do not happen every new Moon and full Moon because the Moon's orbit is tilted at 5 degrees to the ecliptic (p. 9). Eclipses can only take place during an eclipse season (↓).
eclipse (v).

lunar eclipse an eclipse in which the Moon enters the shadow of the Earth. A lunar eclipse is visible from any place on Earth where the Moon is above the horizon. When the Moon lies within the umbra (p. 41) of the Earth's shadow, the eclipse is total; if the Moon lies only partly within the umbra, the eclipse is partial. If the Moon lies within the penumbra (p. 41) of the Earth's shadow, the eclipse is said to be penumbral.

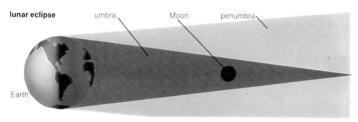

lunar eclipse

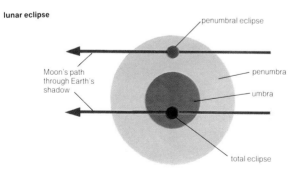

lunar eclipse

total solar eclipse

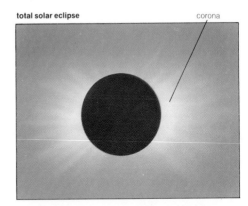

corona

solar eclipse an eclipse in which the Moon moves across the front of the Sun, so that the Moon's shadow falls on the Earth. A solar eclipse is visible only from within the area on which the shadow falls. Within the umbra (p. 41) a total eclipse (p. 40) is seen; in the penumbra (p. 41) a partial eclipse (p. 41) is seen.

eclipse season the times during which the Sun lies close enough to a node (p. 20) of the Moon's orbit so that eclipses can take place. There are two eclipse seasons in an eclipse year (↓).

eclipse year the time taken for the Sun to return to the same node (p. 20) of the Moon's orbit. An eclipse year lasts 346.62 days. It is less than a sidereal year (p. 54) because the nodes of the Moon's orbit regress (p. 20) by about 19 degrees per year.

ecliptic limit the greatest distance that the Sun and Moon can be from the nodes (p. 20) of the Moon's orbit and still cause an eclipse. For a lunar eclipse (↑), the Moon must be within 24 degrees of its node; for a solar eclipse (↑), the Sun must be within 37 degrees of the node. Because the Sun travels along the ecliptic (p. 9) at about 1 degree per day, the eclipse season (↑) for lunar eclipses lasts about 24 days, and the eclipse season for solar eclipses lasts about 37 days.

40 · TIDES AND ECLIPSES/ECLIPSES

Saros (n) the period of 6,585.32 days (just over 18 years) after which the Sun and Moon return to very nearly the same positions in the sky as seen from Earth, and eclipses repeat themselves. There are 223 synodic months (p. 55) in a Saros, and the Saros is less than half a day less than 19 eclipse years (p. 39). The Saros period, which has been known for thousands of years, is an easy way of predicting future eclipses.

total eclipse the condition when light from one body is completely blocked off by another body. A total eclipse of the Sun happens when the Moon completely covers the Sun. A total eclipse of the Moon happens when the Moon lies within the umbra (↓) of the Earth's shadow.

totality (n) the time during which an eclipse is total (↑).

total eclipse of the Sun

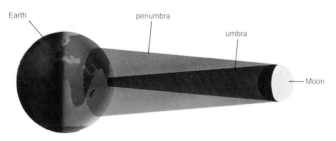

annular eclipse of the Sun

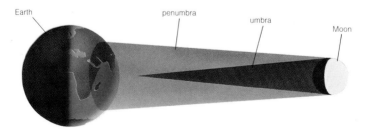

TIDES AND ECLIPSES/ECLIPSES · 41

partial eclipse of the Sun

annular eclipse of the Sun

partial eclipse the condition when only part of the light from one body is cut off by another body. The Sun is partially eclipsed when part of the Moon comes in front of it; the Moon is said to be partially eclipsed when any part of it lies within the umbra (↓) of the Earth's shadow.

penumbral eclipse the condition when the Moon lies in the penumbra (↓) of the Earth's shadow, and part of the Sun's light is cut off from it.

annular eclipse a solar eclipse (p. 39) in which a ring of sunlight remains visible around the edge of the Moon, even when the Moon is in front of the Sun. This happens when the Moon is near apogee (p. 18) so that it appears at its smallest.

umbra[1] (n) the dark centre part of a shadow. People within the umbra of the Moon's shadow see a total eclipse (↑) of the Sun. When the Moon lies within the umbra of the Earth's shadow, it is totally eclipsed.

penumbra[1] (n) the lighter, outer part of a shadow. People within the penumbra of the Moon's shadow see a partial eclipse (↑) of the Sun. When the Moon is within the penumbra of the Earth's shadow, it is said to be in a penumbral eclipse (↑). **penumbral** (adj).

first contact when an eclipse begins. At a solar eclipse (p. 39), it is when the Moon first touches the Sun's edge. At a lunar eclipse (p. 38), it is when the Moon enters the Earth's umbra (↑).

second contact the moment at which an eclipse becomes total. At a solar eclipse (p. 39), it is when the Moon completely covers the face of the Sun. At a lunar eclipse (p. 38), it is when the Moon completely enters the Earth's umbra (↑).

third contact the moment when a total eclipse (↑) ends. At a solar eclipse (p. 39), it is when the Sun begins to reappear from behind the Moon. At a lunar eclipse (p. 38), it is when the Moon begins to move out from the Earth's umbra (↑).

fourth contact the moment when an eclipse ends. At a solar eclipse (p. 39), the face of the Sun is completely uncovered at fourth contact. In a lunar eclipse (p. 38), fourth contact is when all the Moon has left the Earth's umbra (↑).

last contact = fourth contact (↑).

Baily's beads a chain of bright points of light that appears around the edge of the Moon just before and just after a total solar eclipse (p. 39). The beads are caused by sunlight shining between mountains at the Moon's edge.

diamond ring an effect seen at a total solar eclipse (p. 39) just before the last part of the Sun is covered by the Moon before totality (p. 40), or when the first part of the Sun reappears after totality. The diamond ring effect is caused when one of Baily's beads (↑) is much brighter than the rest.

diamond ring effect

occultation
in **b** star is occulted by planet

occultation (n) an event in which a celestial body passes between us and another body so as to block our view of it, e.g. when the Moon, a planet, or an asteroid moves in front of a star. A solar eclipse (p. 39) is really an occultation of the Sun by the Moon. **occult** (v).

grazing occultation an occultation (↑) of a star by the Moon in which the star being occulted just seems to touch the limb (p. 26) of the Moon.

transit

transit² (n) an event in which an object moves across the face of a larger body without occulting (↑) it, e.g. when Mercury or Venus pass in front of the Sun or when a moon or its shadow passes across the face of a planet. **transit** (v).

immersion (n) the moment at which a celestial body enters another body's shadow at an eclipse, or when it moves behind another body at an occultation (↑).

emersion (n) the moment at which a celestial body reappears after an eclipse or an occultation (↑).

appulse (n) the event when two celestial bodies appear to pass close to each other, but without causing an eclipse or occultation (↑).

EARTH AND ITS ENVIRONMENT/GENERAL · 43

Earth

Earth (*n*) the planet on which we live, third in line from the Sun. The Earth has a diameter at its equator of 12,756 km. Its diameter from pole to pole is slightly less, 12,714 km. The Earth orbits the Sun every year at an average distance of 149,600,000 km. It spins on its axis every day. The Earth has one natural satellite (p. 24), the Moon.

terrestrial (*adj*) concerned with, relating to, or similar to the Earth.

telluric (*adj*) of terrestrial (↑) origin.

geo- (*prefix*) relating to the Earth.

geoid (*n*) the shape that the Earth would have if all of it were covered by water. The geoid is not a sphere, because the Earth is oblate (p. 26).

geodesy (*n*) the study of the shape of the Earth and its gravitational field (p. 112). **geodetic** (*adj*).

atmosphere (*n*) the gases around the Earth or other celestial body. The atmosphere of the Earth is made up of about 80 per cent nitrogen and 20 per cent oxygen, plus small amounts of other gases. The atmosphere thins out with height above the Earth, eventually giving way to space.

atmosphere

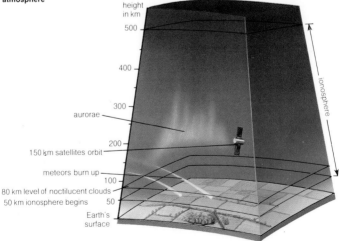

ionosphere (*n*) a part of the Earth's atmosphere from about 50 km to 500 km high in which the atoms (p. 116) are ionized (p. 116) by radiation of short wavelengths (p. 119) (i.e. ultraviolet (p. 120) and X-rays (p. 120)) from the Sun. **ionospheric** (*adj*).

exosphere (*n*) the outermost part of the Earth's atmosphere, from which atoms (p. 116) of gas can escape into space. It starts about 500 km above the Earth's surface and extends out to the magnetopause (↓). This part of the atmosphere is also known as the magnetosphere (↓).

magnetosphere (*n*) the region above the ionosphere (↑), in which the Earth's magnetic field forms a magnetic shell around our planet. It can be thought of as being the outermost region of the Earth's atmosphere. Other planets with magnetic fields also have magnetospheres. It contains the Van Allen belts (↓).

magnetosphere

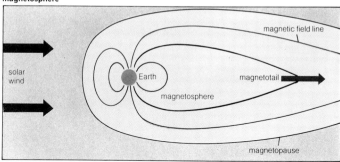

magnetopause (*n*) the outer edge of the magnetosphere (↑), where the magnetosphere gives way to the solar wind (p. 32). On the side facing the Sun the magnetopause lies about 60,000 km from Earth. On the side away from the Sun, it stretches out to form the magnetotail (↓), millions of kilometres long.

magnetotail (*n*) that part of the magnetosphere (↑) that stretches away from the Earth on the side opposite to the Sun. *See also* magnetopause (↑).

Van Allen belts

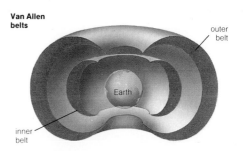

Van Allen belt either of two rings of atomic (p. 116) particles around the Earth, discovered by the American scientist James Van Allen in 1958. The inner belt, which contains protons (p. 116) and electrons (p. 116), lies between 1,000 km and 5,000 km above the Earth. The outer belt, which contains mostly electrons, lies between 15,000 km and 25,000 km above the Earth's equator, but is lower around the magnetic poles (p. 46). The Van Allen belts consist of atomic particles from the Earth's upper atmosphere and from the solar wind (p. 32) trapped by the magnetosphere (↑).

aurora

aurora (*n*) a colourful glow at heights between 100 km and several hundred kilometres above the Earth, caused by the effect of atomic (p. 116) particles from the Sun hitting the gases of the ionosphere (↑). The atomic particles enter the Earth's magnetosphere (↑) and strike the upper atmosphere around the Earth's magnetic poles (p. 46). The brightest aurorae usually happen after a flare (p. 31) on the Sun has thrown out a stream of atomic particles. Some aurorae take the form of arches, while others look like folded curtains. They can be red, yellow or green in colour. **aurorae** (*pl*), **auroral** (*adj*).

auroral oval the ring shape that an aurora (↑) makes around the Earth's magnetic poles (p. 46).
aurora borealis an aurora (↑) in the northern sky.
northern lights = aurora borealis (↑).
aurora australis an aurora (↑) in the southern sky.
southern lights = aurora australis (↑).

magnetic pole either of the two points towards which the needle of a magnetic compass points. The Earth's magnetic poles lie about 11 degrees from the poles of the Earth's rotation, because the axis of the Earth's magnetic field is inclined at 11 degrees to the axis of rotation.

airglow (*n*) a faint light given out by the gases of the upper atmosphere. During the day, molecules (p. 116) are broken up and ionized (p. 116) by ultraviolet (p. 120) light from the Sun. At night the molecules join up again, giving out light as they do so. The airglow occurs in the lower part of the ionosphere (p. 44), from a height of about 90 km upwards.

nightglow (*n*) the airglow (↑) seen at night.

noctilucent cloud a silvery cloud lit up by the Sun long after sunset. It occurs at a height of 80 km, and probably consists of fine dust from burned-up meteors and volcanic eruptions, coated with ice. The clouds are best seen on summer nights in latitudes far from the equator.

night sky light a faint glow of the sky on even the darkest of nights. It is caused mostly by light from the many faint stars, plus the airglow (↑) and zodiacal light (↓).

zodiacal light a faint source of light, caused by sunlight being scattered by small particles of dust in the plane of the ecliptic (p. 9). The zodiacal light requires clear, dark skies to be visible, and is best seen from tropical regions of the Earth. It appears as a cone of light in the west after sunset or in the east before sunrise.

gegenschein (*n*) a rounded patch of very faint light about 10 degrees across, appearing in the sky in the opposite direction from the Sun. Like the zodiacal light (↑), it is caused by sunlight reflected (p. 123) from dust particles in the plane of the ecliptic (p. 9).

scattering (*n*) the reflection (p. 123) of light and other forms of electromagnetic radiation (p. 118) in all directions by dust and gas particles, e.g. the Earth's atmosphere scatters sunlight. Blue light, being of shorter wavelength (p. 119) is scattered more than red light, so that the sky appears blue. **scatter** (*v*).

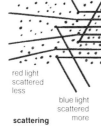

red light scattered less

blue light scattered more

scattering

scintillation (n) the rapid flashing or twinkling of a star, caused by rising and falling currents of air in the Earth's atmosphere which bend light from the star. The solar wind (p. 32) and the ionosphere (p. 44) have similar effects on radio waves (p. 120). **scintillate** (v).

seeing (n) the steadiness of the atmosphere. When the atmosphere is unsteady, stars appear to twinkle (see scintillation ↑) and the image (p. 123) in a telescope jumps around; astronomers call this *bad seeing*. When the air is steady, the image in a telescope is clear and sharp; astronomers call this *good seeing*.

sunset (n) the moment when all the Sun has disappeared below the horizon.

sunrise (n) the moment when the first part of the Sun appears above the horizon.

twilight (n) the time in the evening after sunset, and in the morning before sunrise, when the sky is light even though the Sun is below the horizon. Twilight is caused by the scattering (↑) of sunlight in the Earth's atmosphere.

atmospheric extinction the dimming of starlight caused by scattering (↑) in the Earth's atmosphere. Extinction is most noticeable when a star is close to the horizon, so that its light has to pass through more of the atmosphere. Extinction makes stars appear redder, because red light is scattered less than blue light.

space (n) the region beyond the Earth's atmosphere. Space is empty except for some thinly spread gas and dust; it is very nearly a perfect vacuum (p. 156). The Earth's atmosphere thins out with height, but there is no clear-cut line where the atmosphere ends and space begins. Below about 120 km the atmosphere is too dense for satellites to orbit the Earth, so that for practical purposes a height of 120 km can be said to be the beginning of space.

atmospheric extinction

starlight dimmed by atmosphere

Earth

atmosphere

48 · TIME AND TIMEKEEPING/DAYS

day (*n*) the time taken for the Earth to rotate once on its axis with respect to some fixed point; the word is also used for the rotation period of any planet. There are two sorts of day: the sidereal day (↓) and the solar day (↓). The solar day is the one used for all normal purposes. Astronomers use the sidereal day. **daily** (*adj*).

diurnal (*adj*) daily.

solar day the time taken for the Earth to rotate once on its axis with respect to the Sun. A solar day is divided into 24 hours. The solar day is nearly four minutes longer than the sidereal day (↓) because the Sun appears to move about 1 degree per day against the star background as the Earth goes around its orbit. *See also* solar time (↓).

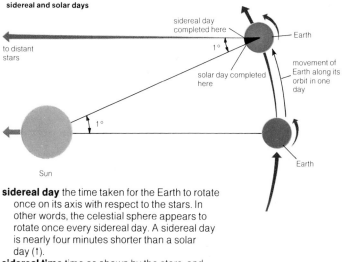

sidereal and solar days

sidereal day the time taken for the Earth to rotate once on its axis with respect to the stars. In other words, the celestial sphere appears to rotate once every sidereal day. A sidereal day is nearly four minutes shorter than a solar day (↑).

sidereal time time as shown by the stars, and measured in sidereal days (↑).

sidereal (*adj*) concerning the stars, of things measured relative to the stars, e.g. the time taken for a planet to complete one orbit or rotate on its axis relative to the stars.

solar time time with respect to the Sun, measured in solar days (↑). There are two sorts of solar time: apparent solar time (↓) and mean solar time (↓).

apparent solar time time given by the daily movement of the Sun across the sky, as shown on a sundial (p. 56). Apparent solar time does not run smoothly and regularly, because the Earth's orbit is elliptical (p. 16) in shape, and also because the Sun moves along the ecliptic (p. 9), not the celestial equator (p. 8). Clocks are set to keep mean solar time (↓).

mean Sun an imaginary body used for the purposes of timekeeping. The mean Sun is imagined as moving along the celestial equator (p. 8) at a steady speed, as would the real Sun if the Earth's orbit were circular (rather than elliptical (p. 16)) and if its axis were not inclined (p. 19). A solar day (↑), as shown by clocks, is actually measured with respect to the mean Sun rather than the real Sun.

mean solar time time with respect to the mean Sun (↑). Mean solar time runs at a steady speed, unlike apparent solar time (↑), and is the time shown by clocks.

mean time = mean solar time (↑).

Greenwich Mean Time GMT. Mean solar time (↑) on the Greenwich meridian (↓). It is also known as Universal Time (↓).

Greenwich meridian the 0 degree line of longitude that passes through Greenwich, London.

Universal Time UT. The name by which Greenwich Mean Time (↑) became known by international agreement in 1928.

Greenwich meridian

equation of time

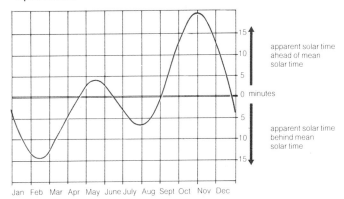

- apparent solar time ahead of mean solar time
- apparent solar time behind mean solar time

equation of time the difference between mean solar time (p. 49) and apparent solar time (p. 49). The difference is greatest in early November when mean solar time is over 16 minutes behind apparent solar time, and in mid February when mean solar time is over 14 minutes ahead of apparent solar time. The two are equal four times a year, on April 15, June 14, September 1 and December 25.

analemma (n) a figure-of-eight shaped curve that shows the difference between apparent solar time (p. 49) and mean solar time (p. 49) throughout the year. The analemma is sometimes marked on globes of the Earth.

time zone any one of twenty-four areas into which the Earth is divided for timekeeping. Each time zone is centred on a line of longitude 15 degrees away from the one next to it. Time in each zone is one hour different from the one next to it, because the Earth rotates through 15 degrees in each hour. The system starts at the zone centred on the line of 0 degrees longitude (the Greenwich meridian (p. 49)). Time moves forward by an hour for each zone to the east of Greenwich, and back by an hour for each zone to the west, until a day is gained or lost at the International Date Line (↓).

time zones

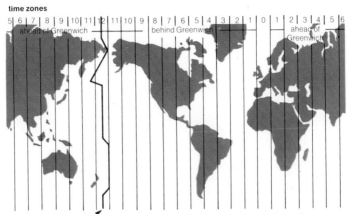

International Date Line

International Date Line an imaginary line passing through the Pacific Ocean, the opposite of the Greenwich meridian (p. 49). The International Date Line lies in the time zone (↑) in which time is 12 hours different from Greenwich. To the east of the Date Line, the date is one day behind the west of the line, e.g. when it is Sunday to the east of the line, to the west it is Monday. The International Date Line does not run exactly along the line of 180 degrees longitude, but bends from side to side where necessary to avoid cutting across land.

ephemeris time ET. A smooth, steady form of time used until 1984 by astronomers when making calculations of the orbital motion of objects in the solar system. Ephemeris time was based on the Earth's orbital motion around the Sun (see tropical year p. 53) rather than on the Earth's rotation on its axis. Ephemeris time was replaced by dynamical time (↓) in 1984.

dynamical time TD. A smooth, steady form of time that replaced ephemeris time (↑) for calculations in 1984. Dynamical time is moving ahead of Universal Time (p. 49) by about one second a year because the Earth's rotation is being slowed down by tidal friction (p. 37) and there are other slight irregularities. See also atomic time (p. 52).

atomic time the world's official time system, in use since 1972, given by the readings of atomic clocks (↓). One second of atomic time equals one second of dynamical time (p. 51). Atomic time is kept in step with Universal Time (p. 49) by adding leap seconds (↓).

atomic clock a clock that uses oscillations of atoms (p. 116) to keep time. One second of atomic time is obtained by counting 9,192,631,770 oscillations of the caesium-133 atom. Atomic clocks run at a constant rate, and they are the most exact form of timekeeping that we have. The best atomic clocks gain or lose no more than 1 second in 100,000 years.

leap second an extra second of time that has to be added to atomic time (↑) about once a year to keep it in step with Universal Time (p. 49). Whereas atomic time is constant, Universal Time is based on the rotation of the Earth, which is slowing down. The leap second is added so that atomic time never differs from Universal Time by more than one second.

calendar (*n*) a timetable for keeping track of the days and months of the year.

lunar calendar a calendar (↑) that is based on the cycle of phases (p. 33) of the Moon, in which 12 synodic months (p. 55) make a lunar year (p. 54). The Muslim calendar is a lunar calendar.

lunisolar calendar a lunar calendar (↑) with an extra month added as necessary to keep the calendar (↑) roughly in step with the seasons, e.g. the Jewish, Hindu and Chinese calendars.

Metonic cycle the period of exactly 19 tropical years (↓) after which the Moon's phases (p. 33) repeat themselves on the same day of the year. There are 235 synodic months (p. 55) in 19 tropical years. The cycle, discovered by the Greek astronomer Meton, was used for making lunisolar calendars (↑).

solar calendar a calendar (↑) based on the yearly motion of the Earth around the Sun, the tropical year (↓), e.g. the Gregorian calendar (↓). Since the tropical year does not contain an exact number of days, every so often an extra day has to be added; *see* leap year (↓).

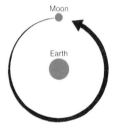

lunar calendar

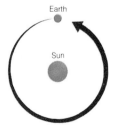

solar calendar

Julian calendar a solar calendar (↑) that was started in 46 BC by the Roman emperor Julius Caesar. It consisted of 365 days per year, with an extra day added every four years. This gave an average length of year of 365.25 days, slightly longer than the actual length of the tropical year (↓). Therefore the Julian calendar slowly got out of step with the seasons.

Gregorian calendar the calendar (↑) used for most purposes throughout the world, started by Pope Gregory XIII in 1582. In the Gregorian calendar an extra day is added at the end of February every four years, but not in century years unless they can be divided by 400, unlike the Julian calendar (↑). So 2000 is a leap year (↓), but 1900 and 2100 are not. This gives an average length of year very close to that of the tropical year (↓).

leap year a year in which an extra day is added to keep the calendar (↑) in step with the seasons.

Julian date JD. A system of dating used by astronomers that simply counts the number of days that have passed since a given starting date. The start point for the Julian day system is chosen as noon on January 1, 4713 BC. The Julian day system provides a continuous time reference that is not affected by changes in the calendar (↑), and it is useful when dealing with events that happen over long periods of time.

year (n) the time taken for the Earth to orbit once around the Sun with respect to a given point. The word is also used for the orbital period (p. 20) of any planet. There are several sorts of year: the tropical year (↓), the sidereal year (p. 54) and the anomalistic year (p. 54).

tropical year the time taken for the Sun to pass once around the sky from one vernal equinox (p. 9) to the next. The tropical year lasts 365.2422 days. Since the position of the Sun determines the occurrence of the seasons, the tropical year is the year on which we base our calendar (↑). The tropical year is slightly shorter than the sidereal year (p. 54) because precession (p. 13) moves the position of the vernal equinox against the star background.

Besselian year a form of the year used by astronomers, equal in length to a tropical year (p. 53). The Besselian year begins and ends when the centre of the mean Sun (p. 49) reaches a right ascension (p. 8) of 18 h 40 min. There is nothing special about this position, other than that it is a round number and that the mean Sun reaches it at some time on January 1 each year.

solar year = a tropical year (p. 53).

sidereal year the time taken for the Earth to orbit the Sun once with respect to the star background, i.e. 365.2564 days.

anomalistic year the time between one perihelion (p. 17) of the Earth and the next, i.e. 365.2596 days. The anomalistic year is longer than the sidereal year (↑) because perturbations (p. 21) of the Earth caused by the planets move the position of the Earth's perihelion.

lunar year the length of 12 synodic months (↓), i.e. 354 days.

season (*n*) one of the four parts into which the year is divided, i.e. winter, spring, summer and autumn. The seasons result from the fact that the Earth's axis is tilted by 23½ degrees, so that the Sun's greatest altitude (p. 11) above the horizon changes during the year, and hence a given place on Earth receives different amounts of sunshine at different times of the year. In the northern hemisphere, winter begins at the winter solstice (p. 10); spring begins at the vernal equinox (p. 9); summer begins at the summer solstice (p. 10); and autumn begins at the autumnal equinox (p. 9). In the southern hemisphere, the seasons are the opposite of those in the northern hemisphere. **seasonal** (*adj*).

seasons

summer in northern hemisphere
winter in southern hemisphere

Sun

winter in northern hemisphere
summer in southern hemisphere

month (*n*) a length of time based on the Moon's orbit around the Earth. The month used in the Gregorian calendar (p. 53) is an invented division of the year that bears only a rough relation to the Moon's orbital period (p. 20). There are several types of month: the synodic month (↓), the sidereal month (↓), the tropical month (↓), the anomalistic month (↓) and the draconic month (↓) **monthly** (*adj*).

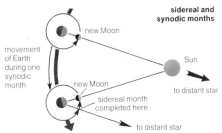

sidereal and synodic months

synodic month the time taken for the Moon to go through a complete set of phases (p. 33), e.g. from one new Moon (p. 33) to the next. The synodic month lasts 29.5306 days. It is the same as a **lunation** (p. 34).

synodic (*adj*) of the time taken for planets and satellites to go once around their orbits measured relative to the Earth, e.g. the time between conjunctions (p. 25) of a planet with the Sun.

sidereal month the time taken for the Moon to complete one orbit of the Earth with respect to the star background. A sidereal month lasts 27.3217 days.

tropical month the time taken for the Moon to complete one orbit of the Earth with respect to the vernal equinox (p. 9). A tropical month lasts 27.3216 days.

anomalistic month the time taken by the Moon to go from one perigee (p. 17) to the next, i.e. 27.5546 days.

draconic month the time from one passage of the Moon through its ascending node (p. 20) to the next, i.e. 27.2122 days.

sundial (*n*) an instrument that shows time during the day by the position of the shadow cast by the Sun. It shows apparent solar time (p. 49).

gnomon (*n*) a stick that casts a shadow, from which the time of day can be read off. In a sundial (↑), the gnomon usually points towards the celestial pole (p. 9).

nocturnal (*n*) an instrument for finding time at night by the position of the stars. The stars of the Great Bear and the Little Bear act as a clock hand as they move around the pole star each night.

transit instrument a telescope used to observe stars as they cross the meridian (p. 10), from which the exact time can be found. The transit instrument is fixed so that it can only point up and down along the observer's meridian.

photographic zenith tube PZT. An instrument used for finding time by the stars. It points directly upwards, and photographs stars as they pass through the zenith (p. 10).

prismatic astrolabe an instrument used for finding time by the stars. It measures the time at which stars reach a certain zenith distance (p. 11), usually an angle of 30 degrees.

PLANETS AND MOONS · 57

Mercury

Venus

dichotomy

relative size of Earth

Mars

Mercury[1] (*n*) the planet closest to the Sun. It is a cratered (p. 34) ball of rock that orbits the Sun every 88 days at an average distance of 58 million km. Mercury is 4,880 km in diameter and spins on its axis every 59 days. It has no moons.

Venus (*n*) the second planet in line from the Sun. Its diameter is 12,100 km and it has a dense atmosphere that consists mostly of carbon dioxide gas. Venus orbits the Sun every 225 days at an average distance of 108 million km. It spins on its axis every 243 days in a retrograde (p. 14) direction. Venus has no moons.

Cytherean (*adj*) of Venus.

dichotomy (*n*) the half-full phase (p. 33) of Mercury or Venus, or of a moon.

ashen light a faint light seen on the night side of Venus when it is at crescent (p. 34) phase (p. 33). Its cause may be similar to that of the Earth's airglow (p. 46).

greenhouse effect the trapping of heat by the gases of a planet's atmosphere. On Venus the greenhouse effect is particularly strong, because the planet's atmosphere consists mostly of carbon dioxide, which holds in the heat very well. Because of the greenhouse effect, the temperature of the atmosphere of Venus is about 475°C. The greenhouse effect also works in the atmosphere of the Earth and of other planets, but not as strongly as on Venus.

Mars (*n*) the fourth planet from the Sun. It is a rocky body, 6,800 km in diameter, and it orbits the Sun every 687 days at an average distance of 228 million km. Mars spins on its axis in 24 h 37.4 min. It has a thin atmosphere that consists mainly of carbon dioxide gas. Mars has two moons, Phobos and Deimos (*see* appendix p. 197). **Martian** (*adj*).

canals of Mars long, straight lines on Mars reportedly seen by some observers in the past and which were believed to be waterways dug by supposed people on Mars. It is now known that the observers were mistaken. There are no canals on Mars, nor is there any sign of life on the planet.

Jupiter (*n*) the largest planet in the solar system, and the fifth in line from the Sun. Jupiter weighs over twice as much as all the other planets put together. It consists mostly of hydrogen and helium. Jupiter is 142,800 km in diameter at its equator, and 134,200 km from pole to pole. It spins on its axis in 9 h 50 min at its equator. Jupiter orbits the Sun every 11.9 years at an average distance of 778 million km. *See* appendix p. 197. **Jovian** (*adj*).

Jupiter

red spot

relative size of Earth

red spot an egg-shaped feature in the clouds of Jupiter, about 32,000 km long and 13,000 km wide. The spot was first seen through telescopes over 300 years ago, and is the only feature in the clouds of Jupiter to have lasted so long. It varies from pale pink to strong orange-red. The spot is thought to be produced by warm gases rising from below the clouds of Jupiter.

Galilean satellite any of the four largest moons of Jupiter – Io, Europa, Ganymede and Callisto – discovered by Galileo in 1610.

Saturn (*n*) the sixth planet from the Sun, with bright rings (↓). Saturn is a ball of gas 120,000 km in diameter across the equator and 108,000 km from pole to pole. It spins on its axis every 10 h 14 min at its equator. Saturn orbits the Sun every 29.5 years at an average distance of 1,430 million km. *See* appendix p. 197.

Saturn

rings of Saturn a flat, circular sheet of material around the equator of Saturn. The rings are made up of countless pieces of rock coated with ice. The pieces range in size from a few millimetres to many metres in diameter. Each piece moves in an orbit around Saturn like a tiny moon. There are three main parts to the rings. The outer part is called *ring A*. The middle part, called *ring B*, is the brightest. The innermost part is *ring C*, also called the crepe ring (because it can be seen through). The overall diameter of the rings is about 270,000 km. The rings of Saturn are either the remains of a former moon that broke up, or are the building blocks of a moon that never formed.

rings of Saturn

Cassini's division a gap about 3,000 km wide between Saturn's ring A and ring B (↑).

PLANETS AND MOONS · 59

Encke's division a narrow gap in Saturn's ring A (↑).
ansa (*n*) the edge of Saturn's rings (↑) as seen from Earth. The edges appear to stick out like handles on each side of the planet. **ansae** (*pl.*).
Roche's limit the closest that a moon can come to a planet without being broken up by the planet's gravitational (p. 112) force. It lies at about 2½ times the radius of the planet from the planet's centre. The rings of Saturn (↑) lie within Roche's limit. Very small objects such as artificial satellites (p. 152) are strong enough to exist within Roche's limit without being torn apart.
Uranus (*n*) the seventh planet, discovered by William Herschel in 1781. Uranus orbits the Sun every 84 years at an average distance of 2,870 million km. It is a ball of gas 52,000 km in diameter. Uranus rotates once every 17.24 hours. The axis of rotation of Uranus is tilted at 98 degrees from the upright, so that it lies almost in the plane of its orbit. Uranus has a set of thin, faint rings around its equator. For details of its five main moons, *see* appendix p. 197.
Neptune (*n*) the eighth planet, discovered by J.G. Galle in 1846. Neptune orbits the Sun every 165 years at an average distance of 4,500 million km. Neptune is a ball of gas 48,000 km in diameter. It spins on its axis about every 18 hours. Neptune has two known moons (*see* appendix p. 197).
Pluto (*n*) the smallest planet in the solar system with a diameter of about 3,000 km. It was discovered by Clyde Tombaugh in 1930. Pluto is a low-density ball of rock and ice, and rotates on its axis every 6 days 9 hours. It orbits the Sun every 250 years at an average distance of 5,900 million km. Pluto is on average further from the Sun than any other planet, but its orbit is so elliptical (p. 16) that at times it comes closer to the Sun than Neptune, as is the case between 1979 and 1999. Pluto has one moon, Charon, (*see* appendix p. 197).
Planet X the name given to a supposed tenth planet that some astronomers think might lie further from the Sun than Pluto. No such planet has yet been seen.

Uranus

relative size of Earth

Neptune

Pluto
·

60 · COMETS

comet (*n*) a body of gas, dust and rock moving on a long, elliptical (p. 16) orbit around the Sun. When a comet is far from the Sun, its gases are frozen into ice; it then shines only by reflecting sunlight. When it nears the Sun, the comet warms up and gas and dust escapes, forming a tail (↓). The gases of the comet become ionized (p. 116) and they give off light of their own. Most comets take many thousands of years to go once around the Sun, but others are seen more often; these are known as periodic comets (↓). Over 1,000 comets have had their orbits calculated. About a dozen new comets are discovered each year. An average comet has a mass about one billionth that of the Earth.

head (*n*) the main part of a comet, consisting of the coma (↓) and nucleus (↓).

coma² (*n*) the region of glowing gas at the head (↑) of a comet, around the nucleus (↓). The coma of a comet is usually about 10,000 km to 100,000 km in diameter.

nucleus² (*n*) the centre of a comet's head (↑), consisting of ice and rock. The nucleus is the only solid part of a comet. The nucleus gives off gas and dust, producing the comet's coma (↑) and tail (↓). A comet's nucleus is about ten kilometres in diameter. The nucleus is covered with a thin crust of dark dust a few centimetres deep.

tail (*n*) a long stream of gas and dust from the head (↑) of a comet. A comet's tail can be 100 million km long, yet it is so thin that stars can be seen through it. A comet's tail always points away from the Sun.

dust tail the part of a comet's tail (↑) that consists of grains of dust, pushed away from the head (↑) by the radiation pressure (↓) of sunlight.

gas tail the part of a comet's tail (↑) that consists of gas blown away from the head (↑) by the solar wind (p. 32). The gas of a comet's tail glows because it is ionized (p. 116).

dirty snowball the description of a comet's nucleus (↑), which consists of a mixture of dust and frozen gases, mostly water, carbon dioxide, ammonia and methane.

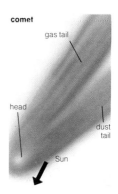

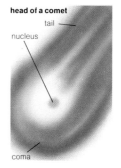

radiation pressure a small force produced by photons (p. 118) hitting a body in space. Radiation pressure of sunlight pushes dust particles away from a comet's head (↑) to form the dust tail (↑).

non-gravitational force the force produced by gas escaping from the nucleus (↑) of a comet, which pushes the comet like a tiny jet engine. Non-gravitational forces cause a comet to wander slightly from its predicted path.

periodic comet a comet that reappears regularly in the inner part of the solar system every 200 years or less.

short-period comet = periodic comet (↑).

comet family a group of comets whose farthest points from the Sun (aphelia p. 18), lie close to the orbits of the giant planets (p. 24), usually Jupiter. Comet families are a result of the gravitational (p. 112) attractions of the planets, which perturb (p. 21) the orbits of comets that pass close to them.

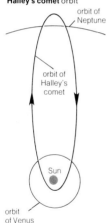

Halley's comet orbit

Halley's comet a large, bright comet that orbits the Sun every 76 years on average, although its orbital period (p. 20) can range from 74 to 79 years. Its orbit was first calculated in 1705 by the English astronomer Edmond Halley. It was first recorded in 239 BC, and last appeared in 1985–86. Its perihelion (p. 17) distance is 0.59 astronomical units (p. 24) and its aphelion (p. 18) is 35 AU. Its next return to perihelion will be in 2061.

Encke's comet the comet with the shortest known period (p. 20), 3.3 years. Its perihelion (p. 17) distance is 0.34 astronomical units and its aphelion (p. 18) is 4.1 AU.

Oort cloud a cloud of 100 billion comets that is thought to lie at the edge of the solar system, about 100,000 astronomical units (p. 24), or 1½ light years (p. 91), from the Sun. The gravitational (p. 112) effect of nearby stars pulls comets out of the cloud and on to orbits that take them towards the Sun, where they become visible. The cloud is named after the Dutch astronomer Jan Oort who put forward the idea of its existence in 1950.

62 · METEORS

meteor (*n*) a quick flash of light in the sky, caused by a particle of dust from space burning up in the Earth's atmosphere. Dust particles, called meteoroids (↓), enter the atmosphere at speeds of 10 km s^{-1} and more. They burn up by friction at a height of about 100 km, producing a line of light that lasts for no more than a second or so. **meteoric** (*adj*).

shooting star = a meteor. Shooting stars are not real stars.

meteoroid (*n*) the name given to any small solid object moving in space. When one burns up in the atmosphere it is known as a meteor; if it is large enough to get through the atmosphere and hit the ground it is called a meteorite.

fireball (*n*) any bright meteor, usually of magnitude (p. 68) −5 or brighter.

bolide (*n*) a meteor that explodes along its path.

meteor stream

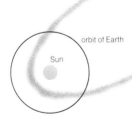

meteor stream a stream of dust particles moving in an orbit around the Sun. The dust particles are left behind by comets, and continue to move along the same orbit as the comet. If the orbit of the meteor stream crosses the orbit of the Earth, a meteor shower (↓) is produced.

meteor shower a large number of meteors that are seen when the Earth meets a meteor stream (↑). All the meteors in a meteor shower appear to come from a point in the sky known as the radiant (↓). Meteor showers occur on certain dates each year *see* table opposite.

radiant

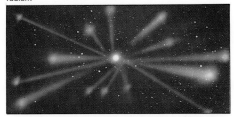

radiant (*n*) the position from which all the meteors in a meteor shower (↑) appear to come. The shower is named after the constellation (p. 95) in which the radiant lies, e.g. the radiant of the Perseid shower lies in Perseus.

meteor showers

shower	date of maximum activity	zenithal hourly rate at maximum
Quadrantids	Jan 3 - 4	100
Lyrids	Apr 21 - 22	15
Eta Aquarids	May 5 - 6	40
Delta Aquarids	July 28 - 29	20
Perseids	Aug 12	60
Orionids	Oct 21	20
Taurids	Nov 3	12
Leonids	Nov 17 - 18	10
Geminids	Dec 13 - 14	60

zenithal hourly rate ZHR. The number of meteors in a meteor shower (↑) that an observer would see each hour in a clear sky if the radiant (↑) were at the zenith (p. 10). If the radiant is not at the zenith, the number of meteors actually seen will be less than the zenithal hourly rate.

sporadic (*n*) a meteor that is not a member of a meteor shower (↑). Sporadic meteors can appear at any time, and appear to move in any direction.

Poynting-Robertson effect an effect in which grains of dust in space slow down in their orbits as a result of being hit by radiation from the Sun. As they slow down they move in towards the Sun, eventually falling into the Sun. However, the very smallest particles of dust and gas are blown away from the Sun by radiation pressure (p. 61).

asteroid (*n*) a body orbiting the Sun that is smaller than a planet. Thousands of asteroids are known, ranging in size from the largest, Ceres (↓), about 1,000 km across, down to those a few hundred metres across. Smaller asteroids probably exist but are too faint to see from Earth. Over 90 per cent of known asteroids lie in the asteroid belt (↓) between Mars and Jupiter. Asteroids are also called minor planets. *See also* Amor asteroids (↓); Apollo asteroids (↓); Trojans (↓).

minor planet = an asteroid.

asteroid belt the part of the solar system between the orbits of Mars and Jupiter where most of the asteroids lie. The asteroid belt is between about 2 and 3.5 astronomical units (p. 24) from the Sun. The asteroids in the belt orbit the Sun about every 3 to 6 years.

Kirkwood gaps a series of spaces in the asteroid belt (↑) where there are few or no asteroids. The Kirkwood gaps lie at exact fractions of the orbital period (p. 20) of Jupiter. The orbital periods of any asteroids in these regions would be commensurable (p. 22) with that of Jupiter. Jupiter's gravity (p. 112) would thus perturb (p. 21) the asteroids on to new orbits, clearing the gaps.

Hirayama families groups of asteroids with similar orbits, believed to have been caused by the break-up of larger asteroids in the past.

Amor asteroid any of a group of asteroids whose perihelion (p. 17) lies between the orbits of Mars and the Earth. Eros (↓) is an example of an Amor asteroid.

Apollo asteroid any of a group of asteroids whose perihelion (p. 17) lies closer to the Sun than the orbit of the Earth. They therefore cross the Earth's path. Hermes (↓) and Icarus (↓) are examples of Apollo asteroids.

Earth-crosser an asteroid that crosses the orbit of the Earth. The Apollo asteroids (↑) are Earth-crossers.

Earth-grazer an asteroid that can come close to the Earth. The Amor (↓) and Apollo (↓) groups of asteroids are Earth grazers.

asteroid belt

orbit of the
Apollo asteroid Icarus

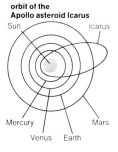

Trojan asteroids

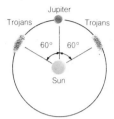

Trojans (*n.pl.*) a group of asteroids that move along the same orbit as Jupiter. They lie at the positions of the Lagrangian points (p. 22) 60 degrees ahead of and behind Jupiter. They are named after the heroes of the Trojan wars.

Ceres (*n*) the largest asteroid, 1,000 km in diameter, and the first to be discovered; it was found by Giuseppe Piazzi in 1801. Ceres orbits the Sun every 4.6 years at an average distance of 2.8 astronomical units (p. 24).

Pallas (*n*) the second asteroid to be discovered, and the second largest, 600 km in diameter. Pallas orbits the Sun every 4.6 years at an average distance of 2.8 astronomical units (p. 24).

Juno (*n*) the third asteroid to be discovered. Juno is 250 km in diameter and orbits the Sun every 4.4 years at an average distance of 2.7 astronomical units (p. 24).

Vesta (*n*) the fourth asteroid to be discovered, and the brightest as seen from Earth. Vesta is 540 km in diameter and orbits the Sun every 3.6 years at an average distance of 2.4 astronomical units (p. 24).

Eros (*n*) a famous Amor asteroid (↑), discovered in 1898. It is cigar-shaped, about 35 km long and 6 km wide. Eros orbits the Sun every 1.8 years on an elliptical (p. 16) path with an aphelion (p. 18) of 1.8 astronomical units (p. 24) and a perihelion (p. 17) of 1.1 astronomical units.

Hermes (*n*) a famous Apollo asteroid (↑) about 1 km in diameter that was discovered in 1937 when it passed 800,000 km from the Earth, closer than any other asteroid. Hermes orbits every 2 years, moving between 0.6 and 2.7 astronomical units (p. 24) from the Sun. It was named after the Greek god, Hermes.

Icarus (*n*) the asteroid with the smallest known perihelion (p. 17) distance, within the orbit of Mercury. Icarus is an Apollo asteroid (↑) that moves in a highly elliptical (p. 16) orbit between 0.19 and 2.0 astronomical units (p. 24) every 1.1 years. It was discovered in 1949, and is about 1.5 km in diameter.

meteorite (*n*) an object from space that hits the surface of the Earth or another body in the solar system. Meteorites are thought to be pieces of asteroids or comets. There are three main types of meteorites: stony meteorites (↓); iron meteorites (↓); and stony-iron meteorites (↓). Most meteorites are slowed down by the Earth's atmosphere so that they drop gently onto the ground. But if a meteorite is moving quickly enough when it hits the ground, it can cause a crater (p. 34). **meteoritic** (*adj*).

micrometeorite (*n*) a tiny grain of dust from space, no more than a few microns (millionths of a metre) across, which does not burn up when it enters the Earth's atmosphere. Micrometeorites fall slowly down through the atmosphere to the Earth's surface.

fall (*n*) a meteorite that is seen to descend to Earth.

find (*n*) a meteorite that is not seen to fall but which is discovered on Earth some time after it has landed.

ablation (*n*) the melting away of the outside of a meteor or spacecraft as it moves at high speed through the atmosphere. **ablate** (*v*), **ablative** (*adj*).

meteorite impact crater in Arizona

impact (*n*) the event when a meteorite hits the ground. If the meteorite is moving at high speed, the force of the impact can produce a large crater (p. 34). Most of the craters on the Moon, Mercury and Mars were produced by meteorite impacts.

stony meteorite the most common type of meteorite observed to fall. Stony meteorites consist mostly of rock, with a small amount of iron. Stony meteorites are divided into two main types: chondrites (↓) and achondrites (↓).

aerolite (*n*) = stony meteorite (↑).

stony meteorite

chondrite (*n*) the most common type of stony meteorite (↑). About 85 per cent of all known meteorites are chondrites. They contain tiny rounded drops known as chondrules (↓).
chondritic (*adj*).

chondrule (*n*) a small, rounded part inside a chondrite (↑) meteorite. It is roughly spherical in shape and measures about 1 mm across. Chondrules are bits of rock that were once molten before they became part of the meteorite.

carbonaceous chondrite a rare form of chondrite (↑) which contains carbon. Carbonaceous chondrites are very soft and break up easily. Some carbonaceous chondrites are believed to have come from the heads of comets.

achondrite (*n*) a type of stony meteorite (↑) that has no chondrules (↑). Unlike the chondrites (↑), achondrite meteorites seem to have been completely melted at some time in the past. They are like some volcanic rocks found on Earth.

iron meteorite

iron meteorite a meteorite made mostly of iron, with about 10 per cent nickel. About 6 per cent of all meteorites seen to fall are irons. But because they do not weather away quickly like stony meteorites, they can still be found long after they have fallen. The largest known meteorite is an iron weighing 60 tons, which lies near Grootfontein in Namibia.

siderite (*n*) = iron meteorite (↑).

Widmanstätten pattern

Widmanstätten pattern a criss-cross pattern of lines seen on the face of an iron meteorite (↑) that has been cut, polished and treated with acid.

stony-iron meteorite

stony-iron meteorite the least common of the three main types of meteorite. About 2 per cent of all meteorites are stony irons. A stony-iron meteorite consists of roughly equal amounts of rock and iron-nickel.

siderolite (*n*) = stony-iron meteorite (↑).

tektite (*n*) a glassy bead a few centimetres across found in certain parts of the world. Tektites are often shaped like teardrops or buttons. They are thought to have been formed by the melting of rock during meteorite impacts (↑) on Earth.

tektites

magnitude (*n*) a measure of the brightness of a celestial body. The brightness that an object appears as seen from Earth is its apparent magnitude (↓); its actual brightness is its absolute magnitude (↓). A difference of one hundred times in brightness between two objects equals a difference of five magnitudes. A difference of one magnitude is therefore equal to the fifth root of 100, which is roughly 2½ times. The bigger the magnitude figure, the fainter the object, i.e. magnitude 2 is fainter than magnitude 1, etc. Objects brighter than magnitude 0 are given negative magnitudes.

apparent magnitude a measure of the brightness that an object has as seen from Earth. This depends not just on the object's actual brightness, but also on its distance. The faintest stars visible to the naked eye on a clear night are of apparent magnitude 6. The faintest objects seen through the largest telescopes are of apparent magnitude 24.

absolute magnitude

10 parsecs

star

absolute magnitude a measure of the actual brightness of a celestial object. Absolute magnitude is the brightness that an object would have if it were at a distance of 10 parsecs (p. 91). For example, the Sun has an apparent magnitude (↑) of –27, but its absolute magnitude is only 4.8. The absolute magnitude is one way for astronomers to compare the luminosity (↓) of stars.

luminosity (*n*) a measure of the total energy given out by a star. The luminosity of a star depends on its size and its temperature, with the biggest and hottest stars being the most luminous. The most luminous stars have the greatest absolute magnitude (↑). *See also* luminosity class (p. 74). **luminous** (*adj*).

visual magnitude the apparent magnitude (↑) of an object as seen by the human eye.

photographic magnitude the apparent magnitude (↑) of a star as measured on a photographic plate that is sensitive to blue light.

photovisual magnitude the apparent magnitude (↑) of a star as measured on a photographic plate that is sensitive to the same range of colours as the human eye.

bolometric magnitude the total amount of energy received from a star at all wavelengths (p. 119), as measured by an instrument known as a bolometer (p. 133).

bolometric correction the difference between the bolometric magnitude (↑) and the visual magnitude (↑) of a star.

extinction (*n*) the dimming of starlight by dust in space. Extinction not only makes stars appear fainter, it also makes them seem redder. Extinction must be taken into account when measuring the magnitude (↑) and colour index (p. 74) of a star.

naked-eye star a star that is visible to the human eye without the use of any optical instrument.

extinction

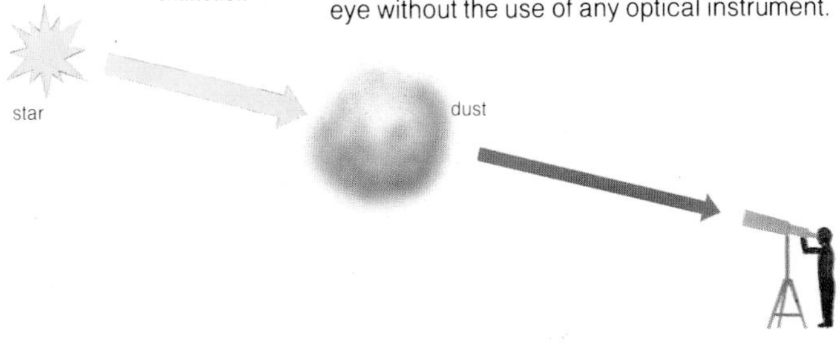

star

dust

star (*n*) a ball of hot gas that gives out light, e.g. our Sun. **stellar** (*adj*).

protostar (*n*) a young star that is being born from the gas of a nebula.

stellar energy the production of energy (heat and light) inside stars by nuclear reactions (p. 70) at their centres. If a ball of gas does not have a mass of at least 7 per cent that of our Sun, conditions of temperature and pressure at its centre are not high enough for nuclear reactions to take place, so the object is not a proper star.

nuclear reaction an event in which an atom (p. 116) of one element (p. 116) is changed into an atom of another element, at high temperature and under strong pressure. Energy is given off in the reaction. Stars produce energy by three main forms of nuclear reaction: the proton-proton chain (↓), the carbon cycle (↓) and the triple alpha process (↓).

proton-proton chain a nuclear reaction (↑) in which hydrogen is changed into helium. Four atoms (p. 116) of hydrogen join to make one atom of helium. The proton-proton chain produces energy in stars up to twice the mass of the Sun, including the Sun itself.

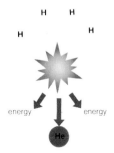
proton-proton chain

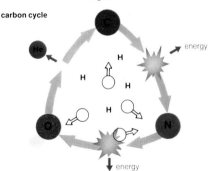
carbon cycle

C = carbon
H = hydrogen
N = nitrogen
O = oxygen
He = helium

carbon cycle a nuclear reaction (↑) which produces energy in stars over twice as massive as the Sun. In the reaction, hydrogen is changed into helium with the help of carbon. Nitrogen and oxygen are also produced, so the reaction is sometimes termed the carbon-nitrogen-oxygen cycle.

carbon-nitrogen-oxygen cycle CNO cycle = carbon cycle (↑).

triple alpha process a nuclear reaction (↑) that occurs inside stars that have used up most of their hydrogen. In the reaction, three atoms (p. 116) of helium join together to form an atom of carbon.

helium flash the event when the triple alpha process (↑) begins in a star, turning helium into carbon.

stellar evolution

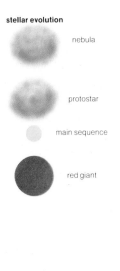

nebula

protostar

main sequence

red giant

stellar evolution

stellar evolution the events in a star's life from birth to old age and eventual death.

Hertzsprung-Russell diagram a graph on which the temperature and brightness of stars is shown. From a star's position on the diagram an astronomer can tell its mass and how far it has reached in its evolution.

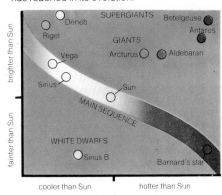

Hertzsprung-Russell diagram

main sequence the part of the Hertzsprung-Russell diagram (↑) in which most stars lie, including the Sun. The position of a star on the main sequence depends on its mass, with the most massive stars at the top left and the smallest stars at the bottom right. A star stays on the main sequence longer than any other part of its lifetime. As stars grow old they move off the main sequence to the upper right on the Hertzsprung-Russell diagram, the region of the red giants (p. 72).

mass-luminosity relation the relationship between the mass and luminosity (p. 68) of stars on the main sequence (↑). The most massive stars are the hottest, and lie at the top end of the main sequence, while the least massive stars are coolest and lie at the bottom end of the main sequence. The mass-luminosity relation does not apply to stars such as red giants (p. 72), which are old stars that have grown away from the main sequence.

72 · STARS/STELLAR EVOLUTION

colour-magnitude diagram a version of the Hertzsprung-Russell diagram (p. 71) in which the colour index (p. 74) of stars is shown against their apparent magnitude (p. 68). Colour-magnitude diagrams are usually made for groups of stars which are all at the same distance from us, in star clusters (p. 88).

dwarf star a star that lies on the main sequence (p. 71) of the Hertzsprung-Russell diagram (p. 71), e.g. the Sun. Dwarf stars get their energy by turning hydrogen into helium by nuclear reactions (p. 70) at their centres. Despite their name, some dwarf stars are actually much larger than the Sun.

red dwarf a small, cool star that lies at the bottom end of the main sequence (p. 71). Red dwarf stars have about one-tenth the mass of the Sun.

sub-dwarf any of a small group of stars that are slightly smaller and fainter than normal dwarf stars (↑). They lie beneath the main sequence (p. 71). They are probably old dwarf stars of Population II (p. 75).

giant star a star that is very much bigger and brighter than the Sun. They are stars that are growing old and reaching the end of their lives. Giant stars lie in the upper right part of the Hertzsprung-Russell diagram (p. 71).

red giant a large, red star at least 10 times the diameter of the Sun. Stars like the Sun swell up into red giants when they grow old.

blue giant a giant star (↑) with a surface temperature of 12,000°C or more, much hotter than a red giant (↑).

sub-giant a large, bright star, smaller than a normal giant (↑). On the Hertzsprung-Russell diagram (p. 71), sub-giants lie between the giants and main sequence (p. 71) stars.

supergiant star the largest and brightest type of star, lying at the very top right part of the Hertzsprung-Russell diagram (p. 71). The most massive stars become supergiants when they grow old.

Hayashi track the path followed by a protostar (p. 69) on to the main sequence (p. 71).

red giant

red dwarf

Hayashi track

spectral type a way of classifying stars into groups according to the nature of their spectrum (p. 119). (*See also* spectral lines p. 121.) The spectral type is a rough guide to the temperature of the star. In order from hottest to coolest, the spectral types are lettered as follows: O,B,A,F,G,K,M. Each spectral type is also subdivided into ten sub-groups from 0 to 9. The range of temperatures, and examples of stars in each class, are shown in the table.

spectral type

spectral type	surface temperature (°C)	examples
O	40,000 - 25,000	Zeta Puppis (supergiant)
B	25,000 - 11,000	Spica (main sequence) Regulus (main sequence) Rigel (supergiant)
A	11,000 - 7,500	Vega (main sequence) Sirius (main sequence) Deneb (supergiant)
F	7,500 - 6,000	Canopus (supergiant) Procyon (subgiant) Polaris (supergiant)
G	6,000 - 5,000	Sun (main sequence) Alpha Centauri (main sequence) Tau Ceti (main sequence) Capella (giant)
K	5,000 - 3,500	Epsilon Eridani (main sequence) Arcturus (giant) Aldebaran (giant)
M	3,500 - 3,000	Barnard's star (main sequence) Antares (supergiant) Betelgeuse (supergiant)

Morgan-Keenan classification a way of classifying stars according to their spectral type (↑) and luminosity class (p. 74).

early type star a hot star of spectral type (↑) O, B or A.

late type star a cool star of spectral type (↑) K or M.

UBV system

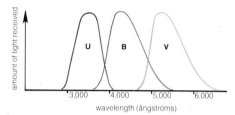

- **colour index** a measure of the colour of a star, and therefore of its temperature. The colour index is found from the difference in brightness of a star measured in two different wavelengths (p. 119), e.g. the blue and yellow parts of the spectrum (p. 119).
- **UBV system** a way of measuring the colour index (↑) of a star from its brightness at ultraviolet (p. 120), blue and yellow wavelengths (p. 119). Three filters are used, each of which lets through light of different wavelengths.

colour index

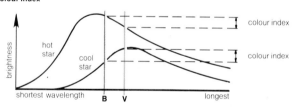

- **luminosity class** a way of classifying stars according to their luminosity (p. 68). The luminosity class of a star shows whether it is a supergiant (p. 72), a giant (p. 72), on the main sequence (p. 71), or a white dwarf (↓). The luminosity classes are given in the table.
- **mass loss** the loss of gas from a star during its lifetime. This can happen in several ways, e.g. in the form of a stellar wind (↓), by throwing off a cloud of gas as in a shell star (p. 84) or a planetary nebula (p. 100), or by an explosion such as a nova (p. 84) or a supernova (p. 76).

luminosity class

Ia	bright supergiant
Iab	supergiant
Ib	less bright supergiant
II	bright giant
III	giant
IV	subgiant
V	main sequence
VI	sub-dwarf
VII	white dwarf

STARS/STELLAR EVOLUTION AND DEATH · 75

Wolf-Rayet stars a small group of very hot stars (surface temperatures above 30,000°C) that are blowing off gas at high speeds, at least 1,000 km s^{-1}. Many Wolf-Rayet stars have companion stars close to them, and as gas flows between the two stars some of it is lost into space. Where Wolf-Rayet stars fit into stellar evolution (p. 71) is not certain.

stellar wind the flow of gas away from a star and into space, like the solar wind (p. 32) from the Sun. In some stars, such as red giants (p. 72), the stellar wind is much stronger than it is from our Sun, and these stars can lose large amounts of mass as a result.

stellar population either of the two different age groups of stars, Population I (↓) and Population II (↓). Population II stars are the oldest, having been born when galaxies first formed. Population I stars formed more recently, and include the Sun. Some stars, known as extreme Population I, are still forming today.

Population I a group of stars that are found in the spiral arms (p. 104) of galaxies. They are relatively young stars, having formed within the past few billion years. The Sun is a Population I star. Population I stars contain more heavy elements (p. 116) than those of Population II (↓).

Population II a group of stars that are found in the centres of galaxies and in globular clusters (p. 88). Population II stars were the first stars to form, more than 10 billion years ago. They contain almost no heavy elements (p. 116).

white dwarf a small, faint star that is the end point in the life of a star like the Sun, when it has run out of fuel for nuclear reactions (p. 70). A white dwarf has a mass about that of the Sun, but a diameter only about that of the Earth, i.e. roughly 1 per cent the diameter of the present-day Sun. The material of a white dwarf is therefore very dense. *See also* planetary nebula (p. 100).

Sirius B the first white dwarf (↑) to be discovered. Sirius B is a companion to Sirius which it orbits every 50 years. Sirius B was first seen in 1862. It has the same mass as the Sun, but is only 0.02 times its diameter.

white dwarf

76 · STARS/STELLAR DEATH

electron degeneracy the condition in a very dense gas when electrons (p. 116) are so tightly packed together that they stop the gas from becoming any denser. Electron degeneracy occurs in white dwarf (p. 75) stars. It balances the force of gravity (p. 112), and prevents the white dwarf from getting any smaller.

degenerate matter a very dense form of matter in which the atomic (p. 116) particles are crowded together as closely as is physically possible. Once the matter in a star has reached the state of degeneracy, the matter cannot be pressed together any more tightly and so the star cannot become any smaller. In a white dwarf (p. 75), the matter is in a state of electron degeneracy (↑). In a neutron star (↓), the degenerate matter is made of neutrons (p. 116).

Chandrasekhar limit the heaviest that a white dwarf (p. 75) can be, i.e. 1.4 solar masses. If a white dwarf has a mass greater than 1.4 solar masses, its gravity (p. 112) is strong enough to overcome the effect of electron degeneracy (↑), and the object then turns into a neutron star (↑) or a black hole (p. 79).

neutron star a dead star, smaller and denser even than a white dwarf (p. 75). A neutron star consists of degenerate matter (↑) made of neutrons (p. 116). The neutrons are formed from protons (p. 116) and electrons (p. 116) that have been forced to join together either by gravity (p. 112) or by the effect of a supernova (↓). Neutron stars have masses up to about three times that of the Sun, but are only about 20 km across, i.e. less than 1 per cent the diameter of a white dwarf. See also pulsars (p. 78).

supernova (n) a star that explodes, increasing in brightness by millions of times for a few weeks or months. A supernova is thousands of times brighter than an ordinary nova (p. 84). The brightest supernovae can be as bright as an entire galaxy. In a supernova explosion, a star throws off most of its mass at high speeds, leaving behind a supernova remnant (↓). There are two main sorts of supernova: Type I (↓) and Type II (↓). **supernovae** (pl).

neutron star compared in size with the island of Malta

STARS/STELLAR DEATH · 77

Type I supernova a class of supernova (↑) in which about one solar mass of gas is thrown off at speeds of around 12,000 km s^{-1}. Type I supernovae are thought to be caused by white dwarf (p. 75) stars that have a close companion star. Gas flowing from the companion onto the white dwarf makes the white dwarf heavier than the Chandrasekhar limit (↑). The white dwarf then explodes.

Type II supernova a class of supernova (↑) in which a massive star explodes at the end of its life. Stars that become Type II supernovae have a mass of at least four times that of the Sun. A Type II supernova thows off several solar masses of gas at speeds of around 5,000 km s^{-1}.

supernova remnant the remains of a star that has exploded as a supernova (↑). The gas thrown out by the supernova moves off into space, forming a shell around the exploded star. The central part of the exploded star can be left behind as a neutron star (↑) or a black hole (p. 79). Supernova remnants give out large amounts of radio waves (p. 120) and X-rays (p. 120). A famous supernova remnant is the Crab nebula (↓).

Crab nebula

Crab nebula a supernova remnant (↑) about 6,000 light years away in the constellation (p. 95) of Taurus the bull. The Crab nebula is the remains of a star that was seen to explode in the year 1054. It is called the Crab nebula because the first observers who saw it through telescopes thought it looked like the shape of a crab. At the centre of the Crab nebula is a pulsar (p. 78).

Cassiopeia A a supernova remnant (↑) about 10,000 light years away in the constellation (p. 95) of Cassiopeia. It is one of the strongest radio sources (p. 134) in the sky.

runaway star a young, hot star moving quickly through space at speeds of 100 km s^{-1} and more. Runaway stars may once have been members of close binary star (p. 85) systems, but were sent into space when the other member exploded as a supernova (↑).

pulsar (*n*) an object that gives out rapid flashes of radiation, usually every second or so. Pulsars are believed to be neutron stars (p. 76) that are spinning rapidly. Each time they spin we receive a beam of radiation from them, like a lighthouse beam. Pulsars were discovered in 1967 by radio astronomers; over 300 radio pulsars are now known. Another class of pulsars gives out energy at X-ray (p. 120) wavelengths (p. 119). X-ray pulsars are members of close double star (p. 85) systems in which gas falls onto the neutron star from a companion star.

Crab pulsar the pulsar (↑) that lies at the centre of the Crab nebula (p. 77). The Crab pulsar is a neutron star (p. 76), the remains of the star that exploded to form the Crab nebula. The Crab pulsar gives out a pulse of energy at radio to gamma-ray (p. 120) wavelengths (p. 119) 30 times every second, making it one of the fastest pulsars known. It is one of only two pulsars seen to flash at visible wavelengths, the other one being the Vela pulsar (↓).

Vela pulsar a pulsar (↑) in the constellation (p. 95) of Vela that pulses 11 times every second. It is only the second pulsar seen to flash at visible wavelengths (p. 119), the first being the Crab pulsar (↑).

binary pulsar the first known example of a radio pulsar (↑) that is in orbit around another star, discovered in 1974. It pulses 17 times a second and orbits its companion star every 7 h 45 min. The companion star is probably also a neutron star (p. 76), but is not a pulsar.

black hole

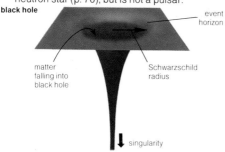

black hole an object whose pull of gravity (p. 112) is so great that nothing can escape, not even light, so the object is truly black. Black holes are thought to be formed when stars many times heavier than the Sun die. If the dead star weighs more than about three solar masses its own gravity is so strong that nothing can stop its gravitational collapse (p. 112). It gets ever smaller and denser, until it becomes a black hole. Although nothing can get out of a black hole, things can fall in. When gas falls towards a black hole it heats up to temperatures of millions of degrees. At such temperatures the gas gives out X-rays (p. 120) that can be picked up by satellites in space. Several objects that are probably black holes have been found in this way, e.g. Cygnus X-1 (↓).

Cygnus X-1

Cygnus X-1 an X-ray (p. 120) source in the constellation (p. 95) of Cygnus that may be the first known example of a black hole (↑). Cygnus X-1 orbits a visible star from which gas falls onto Cygnus X-1, heating up and giving out X-rays. The estimated mass of Cygnus X-1 is 8 solar masses, too heavy for it to be a neutron star (p. 76) so it is probably a black hole.

event horizon the surface of a black hole (↑). Once something has crossed the event horizon it can never again escape. It is trapped for ever inside the black hole.

Schwarzschild radius a measure of the size of a black hole (↑). It is the radius of the event horizon (↑). The more massive the black hole, the larger the Schwarzschild radius.

ergosphere (n) the region just outside the event horizon (↑) of a rotating black hole (↑). Anything that enters the ergosphere is forced to spin around the black hole, as though it were in a whirlpool.

singularity (*n*) a point of infinite density at the centre of a black hole (p. 79). All the mass of the star that formed the black hole, plus anything else that falls in after it, is pressed together into the singularity. There is nothing between the singularity and the event horizon (p. 79) except empty space.

naked singularity a singularity (↑) that does not have an event horizon (p. 79) around it. According to theory, a naked singularity might occur if a black hole (p. 79) were spinning very quickly, but astronomers are not sure whether naked singularities really do exist.

variable stars on the Hertzsprung-Russell diagram

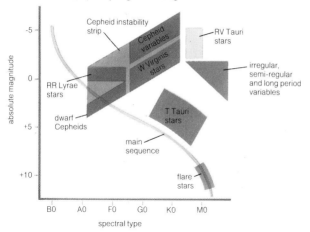

vary (*v*) to change, to become different. **variable** (*adj*), **variation** (*n*).

variable star a star that changes in brightness. Some stars vary because of actual changes in the star itself. Other stars appear to vary because something comes between us and the star, blocking off some or all of the star's light (*see* eclipsing binary p. 85).

period[2] (*n*) the time taken for a variable star (↑) to go from its brightest to its faintest and back again.

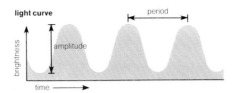

light curve a graph showing how the brightness of a variable star (↑) changes with time.

amplitude (*n*) the range between the brightest and faintest magnitudes (p. 68) of a variable star (↑).

pulsating variable star a star whose changes in brightness are caused by changes in its size. Some pulsating variable stars change regularly, e.g. Cepheid variables (↓) and RR Lyrae variables (p. 82), while others are semi-regular (p. 83) and irregular variables (p. 83).

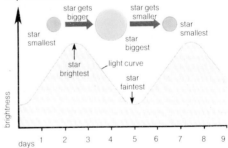

Cepheid variable any of a group of pulsating variable stars (↑) that change regularly in brightness by between about 0.1 and 2 magnitudes (p. 68) every 1 to 100 days. Cepheid variables are bright yellow supergiants (p. 72). The brightest Cepheids take the longest to change in brightness (*see* period-luminosity relation p. 82).

Delta Cephei the first Cepheid variable (↑) star to be discovered, and from which all others take their name. Delta Cephei varies from magnitude (p. 68) 3.5 to 4.4 and back again every 5.37 days.

W Virginis stars a group of Cepheid variables (p. 81) that are found in globular clusters (p. 88). They are about 1.5 magnitudes (p. 68) fainter than ordinary Cepheids that have the same period (p. 80) of variation. W Virginis stars are members of Population II (p. 75).

period-luminosity relation a law which shows that the period (p. 80) of a Cepheid variable (p. 81) is related to its average absolute magnitude (p. 68). The longer a Cepheid's period, the brighter the star is. By measuring the time a Cepheid takes to vary, astronomers can use the period-luminosity relation to work out the star's actual brightness and from this find its distance. Because of this Cepheids are important for finding distances in space.

period-luminosity relation

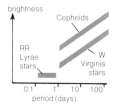

RR Lyrae variable any of a group of regularly pulsating variable stars (p. 81) with periods (p. 80) of less than a day. RR Lyrae variables are old giant stars (p. 72) of Population II (p. 75). They are usually found in globular clusters (p. 88), and so are sometimes called cluster variables. All RR Lyrae variables have the same average absolute magnitude (p. 68) of +0.5.

cluster variable = RR Lyrae variable (↑).

dwarf Cepheid any of a group of pulsating variable stars (p. 81) similar to RR Lyrae variables (↑) but fainter and with shorter periods (p. 80).

Delta Scuti stars a group of pulsating variable stars (p. 81) with periods (p. 80) of a few hours and amplitudes (p. 81) of a few tenths or even a few hundredths of a magnitude (p. 68). They are similar to dwarf Cepheids (↑) but have smaller amplitudes.

Cepheid instability strip a part of the Hertzsprung-Russell diagram (p. 71) in which lie Cepheid variables (p. 81), RR Lyrae stars (↑), dwarf Cepheids (↑), and Delta Scuti stars (↑).

Beta Cephei stars a small group of blue giant (p. 72) pulsating variable stars (p. 81). They have small amplitudes (p. 81) and short periods (p. 80).

Beta Canis Majoris stars = Beta Cephei stars (↑).

long-period variable a pulsating variable star (p. 81) with a period (p. 80) from about 80 days to two years or longer. They have large amplitudes (p. 81) of 2.5 magnitudes (p. 68) and above. They are all red giants (p. 72), and there are more of them than any other type of variable star (p. 80). Unlike Cepheid variables (p. 81), the period and the amplitude of long-period variables is not repeated exactly each time.

Mira (n) the best-known example of a long-period variable (↑), and the first of its kind to be discovered. Mira is a red giant (p. 72) star in the constellation (p. 95) of Cetus. Mira varies by about 5 or 6 magnitudes (p. 68) every 331 days on average. Also known as **Omicron Ceti**.

semi-regular variable a pulsating variable star (p. 81), giant (p. 72) or supergiant (p. 72), whose amplitude (p. 81) and period (p. 80) is only roughly regular. They usually vary by about one or two magnitudes (p. 68) every 100 days or so.

RV Tauri stars a group of semi-regular variables (↑) consisting of yellow supergiants (p. 72) with periods (p. 80) from 30 to 150 days and amplitudes (p. 81) from 1 to 3 magnitudes (p. 68).

irregular variable a pulsating variable star (p. 81) with no fixed period (p. 80) or amplitude (p. 81). Irregular variables are giant (p. 72) and supergiant (p. 72) stars. They usually vary by no more than a magnitude (p. 68) or two.

eruptive variable a star that shows sudden, usually very large (i.e. several magnitudes p. 68), changes in brightness. Eruptive variables include flare stars (↓) and novae (p. 84).

cataclysmic variable = eruptive variable (↑).

flare star a faint red dwarf (p. 72) that brightens without warning by up to 100 times in a few minutes. The brightness increase is thought to be caused by a flare (p. 31) on the surface of the red dwarf. These flares are probably similar to those on the Sun, but are much more noticeable because the star is normally so dim.

UV Ceti a famous flare star (↑). UV Ceti increases in brightness by one or two magnitudes (p. 68) in under a minute, and then fades back to its normal brightness in two or three minutes.

T Tauri stars a peculiar group of variable stars (p. 80) that are surrounded by a cloud of gas and dust from which they are believed to be forming, i.e. they are protostars (p. 69). The variations are thought to be caused as the star grows from the gas around it. T Tauri itself has a spectral type (p. 73) similar to the Sun, and varies irregularly by up to four magnitudes (p. 68).

shell star a hot, bright star, spinning so rapidly that it throws off shells of gas, changing in brightness as it does so.

nova (n) a faint star that suddenly increases in brightness by thousands or tens of thousands of times for a few days or weeks before fading back again. Novae are thought to be binary stars (↓) containing a white dwarf (p. 75) and a normal star. Gas from the normal star flows on to the white dwarf where it causes a small explosive nuclear reaction (p. 70) and gas is thrown off. Unlike a supernova (p. 76), a nova explosion does not destroy the star. **novae** (pl).

nova

normal star

white dwarf

gas flows onto white dwarf

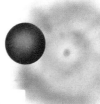
gas thrown off in explosion

recurrent nova a nova (↑) that has been seen to increase in brightness more than once. All novae may undergo more than one outburst.

dwarf nova a faint star that brightens by up to 5 magnitudes (p. 68) every few weeks or months before fading again. The cause of dwarf novae is thought to be similar to that of ordinary novae (↑), except that the brightness increase is less and the outbursts happen more often.

spectrum variable a star with a strong magnetic field. Changes in the strength of the magnetic field every few days or weeks causes slight changes in the star's brightness.

magnetic star = spectrum variable (↑).

eclipsing binary

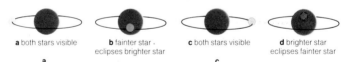

a both stars visible **b** fainter star eclipses brighter star **c** both stars visible **d** brighter star eclipses fainter star

a
light curve

c

d

b

eclipsing binary a double star (↓) in which one star passes in front of the other and cuts off some or all of its light from Earth. The total light received on Earth from the two stars therefore drops for a while, although the brightness of the stars themselves does not change. When one star is in front of the other, that event is known as an eclipse. How often the eclipses happen depends on how long it takes one star to orbit the other.

Algol (n) a famous eclipsing binary (↑) and the first one to be discovered. Every 2.87 days the brighter star of Algol is eclipsed by a fainter companion. During the eclipse the brightness of Algol drops from magnitude (p. 68) 2.2 to 3.5. The eclipse lasts 10 hours.

double star a pair of stars seen close together in the sky. In most cases, the two stars that form a double are actually related, i.e. they lie at the same distance from us and are held together by gravity (p. 112). Such a pair is also known as a physical double (p. 86) star or a binary (↓). But a small percentage of double stars are optical doubles (p. 86).

multiple star a group of more than two related stars.

binary star a physical double (p. 86) star. The two stars in a binary are in orbit around each other. If one star passes in front of the other as seen from Earth, the pair is known as an eclipsing binary (↑).

physical double two stars at the same distance from us (or almost so) and held together by gravity (p. 112), unlike an optical double (↓). Astronomers usually call a physical double star a binary (p. 85).

optical double two stars that lie in the same line of sight as seen from Earth, although one is actually much further away from us than the other. The two stars in an optical double are therefore not related.

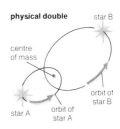

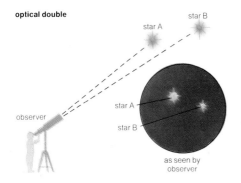

primary[2] (n) the brighter member of a double star (p. 85).

secondary[2] (n) the fainter member of a double star (p. 85).

spectroscopic binary a binary star (p. 85) in which the two stars are so close together that they appear as one through even the most powerful telescope. We can only tell that the star is a binary by studying its light in a spectroscope (p. 132). The spectroscope reveals the Doppler effect (p. 114) in the light as one star moves around the other.

single-line spectroscopic binary a spectroscopic binary (↑) in which one star is so faint that only the light from the brighter star is seen.

mass function a relationship between the masses of the two stars in a single-line spectroscopic binary (↑). It is not possible to calculate the separate masses of each star.

double-line spectroscopic binary a spectroscopic binary (↑) in which both stars are bright enough for their light to be seen.

spectrum binary a double-line spectroscopic binary (↑) in which we see the orbits of the two stars almost directly from above, so there is no Doppler effect (p. 114) in the stars' light.

astrometric binary

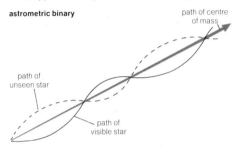

path of centre of mass

path of unseen star

path of visible star

astrometric binary a binary star (p. 85) in which the fainter star cannot be seen directly, but which gives itself away by its effect on the proper motion (p. 89) of the brighter star. The brighter star moves slightly from side to side as it orbits the fainter star, and this side-to-side movement can be measured through telescopes.

X-ray binary a binary star (p. 85) that emits X-rays (p. 120). The X-rays are thought to be caused by hot gas from a normal star falling onto its companion, which can be a white dwarf (p. 75), a neutron star (p. 76) or a black hole (p. 79).

mass transfer the flow of gas from one star in a binary (p. 85) pair to the other. Mass transfer happens when one star has grown big enough to fill its Roche lobe (↓).

Roche lobe a figure-eight shaped volume of space around the stars in a binary (p. 85) system. If one star swells up to fill its Roche lobe, gas can fall from that star onto its companion.

contact binary a close binary star (p. 85), in which both stars fill their Roche lobes (↑).

accretion disk a ring of gas orbiting one member of a binary (p. 85) system, formed by gas flowing from the companion star.

Roche lobe

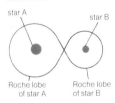

star A — star B
Roche lobe of star A — Roche lobe of star B

accretion disk

star A — star B
flow of gas — accretion disk

star cluster a group of stars lying together in space. All the stars in a given cluster are thought to be born from the same large cloud of gas. Clusters can contain anything from a handful of stars to hundreds of thousands of stars. In some clusters the stars are held together by gravity (p. 112), but in others they are slowly drifting apart.

open cluster a group of stars in the spiral arms (p. 104) of a galaxy. Open clusters have no particular shape, and the stars within them are usually loosely scattered, unlike the tightly packed globular clusters (↓). Open clusters contain up to a few hundred stars of Population I (p. 75).

galactic cluster = open cluster (↑).

moving cluster a group of stars moving through space together in the same direction at similar speeds. The Hyades (↓) is an example of a moving cluster.

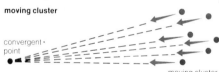

moving cluster

convergent point the point towards which the members of a moving cluster (↑) appear to be moving.

globular cluster a roughly ball-shaped group of hundreds of thousands of densely packed stars. Globular clusters are scattered around galaxies, including our own, and came into being while the galaxies were still forming. They contain old stars of Population II (p. 75).

stellar association a widely spread group of young stars that has recently been born in a spiral arm (p. 104) of our Galaxy, like a very large open cluster (↑).

OB association a stellar association (↑) consisting mainly of hot, bright stars of spectral types (p. 73) O and B.

T association a stellar association (↑) consisting of T Tauri stars (p. 84).

globular cluster

Pleiades star cluster

field star a star that is visible in the same field of view as a star cluster (↑), but is not a real member of the cluster, because it lies at a different distance.

Pleiades (n) a famous open cluster (↑) in the constellation (p. 95) of Taurus, the bull. It is commonly known as the Seven Sisters, because six or seven stars are bright enough to be seen by the naked eye. But the whole cluster contains several hundred stars, all formed within the past 50 million years or less. The Pleiades lie 415 light years (p. 91) away.

Hyades (n) an open cluster (↑) of about 200 stars in the constellation (p. 95) of Taurus, the bull. The Hyades cluster lies 150 light years (p. 91) away.

astrometry (n) the measurement of the positions of stars and other celestial objects on the celestial sphere.

aberration of starlight
raindrops seen from a moving vehicle seem to fall at an angle; this effect is the same as that of the aberration of starlight due to the movement of the Earth

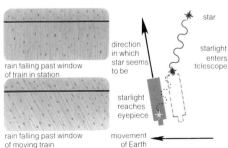

aberration of starlight an effect caused by the movement of the Earth across the path of the rays of light coming from a star. The effect makes the light rays appear to be coming in at an angle, so that the star seems to be in a slightly different position from its real one.

proper motion
the stars of the Plough (Big Dipper) as they appear today

and as they will appear in 100,000 years time because of the effect of proper motion

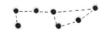

proper motion the movement of a star on the celestial sphere that results from its own movement and also the movement of the Sun as they both orbit the Galaxy; symbol μ. Proper motion causes a change in the position of stars over long periods of time. The star with largest proper motion is Barnard's star (p. 93).

peculiar motion (1) the proper motion (p. 89) of a star with the effects of the Sun's movement removed. The peculiar motion of a star is its own movement through space; (2) the movement of one star relative to other stars near it.

radial velocity the velocity of a star towards or away from the Sun, measured from the Doppler effect (p. 114) on its light. The radial velocity is given as a positive figure if the star is moving away, negative if it is coming towards us.

tangential velocity the velocity of a star at right angles to the line of sight.

transverse velocity = tangential velocity (↑).

radial velocity

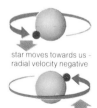

star moves towards us - radial velocity negative

star moves away from us - radial velocity positive

space velocity

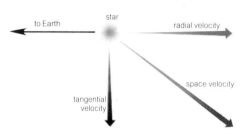

space velocity the velocity of a star relative to the Sun, resulting from its radial velocity (↑) and its tangential velocity (↑).

peculiar velocity the velocity of a star relative to the local standard of rest (↓).

local standard of rest a volume of space around the Sun in which the space velocities (↑) of all the stars average out to zero. Stars out to about 100 parsecs (↓) from the Sun are included in the local standard of rest.

solar motion the movement of the Sun relative to the local standard of rest (↑). The speed of the solar motion is about 20 km s^{-1}. *See also* solar apex (↓).

solar apex the direction in space towards which the Sun is moving relative to the local standard of rest (↑). The solar apex lies on the border between the constellations (p. 95) of Lyra and Hercules, near to the bright star Vega.

parallax

solar antapex the direction in space opposite to the solar apex (↑). The solar antapex lies in the constellation (p. 95) of Columba.

parallax (*n*) the change in position of a star as seen from opposite sides of the Earth's orbit; symbol π. The amount of a star's parallax shows how far away it is, with the nearest stars having the largest parallax. To measure parallax, the star's position is observed twice, six months apart. Beyond about 100 light years (↓) the parallax of stars is too small to be measurable. **parallactic** (*adj*).

trigonometric parallax = parallax (↑).

secular parallax the change in position of stars with time because of the solar motion (↑).

statistical parallax the average distance of a large number of stars, found from their proper motions (p. 89) and radial velocities (↑).

moving cluster method a way of finding the distance of the stars in a cluster that are moving through space together. The distance to the Hyades (p. 89) star cluster (p. 88), measured in this way, is an important first step to finding the distances of many other stars and galaxies.

spectroscopic parallax the distance of a star found from a study of its spectrum (p. 119), i.e. the star's spectral type (p. 73) and luminosity class (p. 74) tell us its absolute magnitude (p. 68). The star's distance can then be calculated from its distance modulus (↓). This is the way in which astronomers find the distances of most stars that are too far away to show a normal parallax (↑).

distance modulus the difference between the apparent magnitude (p. 68) and the absolute magnitude (p. 68) of a star or galaxy, from which its distance can be calculated.

light year the distance travelled by a beam of light in one year, i.e. 9.46×10^{12} km.

parsec (*n*) the distance at which a star would have a parallax (↑) of one second of arc (p. 15). One parsec = 3.2616 light years (↑); symbol pc.

light time the time taken for a beam of light to reach the Earth from a celestial body.

STARS/BRIGHTEST & NEAREST

the brightest stars

name	constellation	apparent magnitude
Sirius	Canis Major	-1.46
Canopus	Carina	-0.72
Alpha Centauri	Centaurus	-0.27
Arcturus	Boötes	-0.04
Vega	Lyra	+0.03
Capella	Auriga	+0.08
Rigel	Orion	+0.12
Procyon	Canis Minor	+0.38
Achernar	Eridanus	+0.46
Betelgeuse	Orion	+0.50 (variable)
Beta Centauri	Centaurus	+0.61
Altair	Aquila	+0.77
Aldebaran	Taurus	+0.85
Alpha Crucis	Crux	+0.87
Antares	Scorpius	+0.96 (variable)
Spica	Virgo	+0.98
Pollux	Gemini	+1.14
Fomalhaut	Piscis Austrinus	+1.16
Deneb	Cygnus	+1.25
Beta Crucis	Crux	+1.25
Regulus	Leo	+1.35

Achernar (n) the ninth-brightest star in the sky, of magnitude (p. 68) 0.5. It is a blue-white star lying 85 light years (p. 91) away in the constellation (p. 95) of Eridanus. Also known as **Alpha Eridani**.

Aldebaran (n) a red giant (p. 72) star in the constellation (p. 95) Taurus. It lies 68 light years (p. 91) away, and appears of magnitude (p. 68) 0.9. Also known as **Alpha Tauri**.

Alpha Centauri the third-brightest star in the sky, and also the closest to the Sun. Alpha Centauri lies 4.3 light years (p. 91) away in the constellation (p. 95) of Centaurus, and shines at magnitude (p. 68) -0.27. Telescopes show it is a group of three stars, one of which, Proxima Centauri (p. 94), is slightly closer to us than the other two. Also known as **Rigil Kentaurus**.

Alpha Crucis a blue-white star of magnitude (p. 68) 0.9, lying 360 light years (p. 91) away in the constellation (p. 95) of Crux, the Southern Cross. Also known as **Acrux**.

Altair (n) a white star 16 light years (p. 91) away in the constellation (p. 95) of Aquila. It appears of magnitude (p. 68) 0.8. Also known as **Alpha Aquilae**.

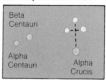

relative positions of stars

relative positions of stars

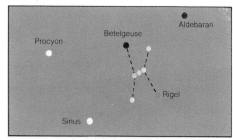

Antares (*n*) a red supergiant star (p. 72) 300 times the diameter of the Sun, 330 light years (p. 91) away in the constellation (p. 95) Scorpius. Its magnitude (p. 68) is about 1.0, but it is slightly variable (p. 80). Also known as **Alpha Scorpii**.

Arcturus (*n*) the fourth-brightest star in the sky, of magnitude (p. 68) -0.04. It is a red giant (p. 72). Arcturus lies 36 light years (p. 91) away in the constellation (p. 95) of Boötes. Also known as **Alpha Boötis**.

Barnard's star the second-closest star to the Sun, lying 6 light years (p. 91) away in the constellation (p. 95) Ophiuchus. It is a 9th-magnitude (p. 68) red dwarf (p. 72). Barnard's star has the largest proper motion (p. 89) of any star known, covering 10 arc seconds (p. 15) per year. It may have planets around it.

Beta Centauri a blue giant (p. 72) star of magnitude (p. 68) 0.6, lying 460 light years (p. 91) away in the constellation (p. 95) Centaurus. Also known as **Hadar**.

Betelgeuse (*n*) a red supergiant star (p. 72) over 300 times the diameter of the Sun, lying 310 light years (p. 91) away in the constellation (p. 95) of Orion. Betelgeuse is a variable star (p. 80), but has an average magnitude (p. 68) of 0.5. Also known as **Alpha Orionis**.

Canopus (*n*) the second-brightest star in the sky, of magnitude (p. 68) -0.7. It is a yellow-white supergiant (p. 72) about 300 light years (p. 68) away in the constellation (p. 95) Carina. Also known as **Alpha Carinae**.

Capella (*n*) the sixth-brightest star in the sky, of magnitude (p. 68) 0.08. It is a yellow star lying 42 light years (p. 91) away in the constellation (↓) of Auriga. Also known as **Alpha Aurigae**.

Lalande 21185 the fourth-closest star to the Sun, lying 8.1 light years (p. 91) away in the constellation (↓) Ursa Major. Lalande 21185 is a red dwarf of magnitude (p. 72) 7.5. It may have a planetary system.

Procyon (*n*) the eighth-brightest star in the sky, of magnitude (p. 68) 0.4. It lies 11.3 light years (p. 91) away in the constellation (↓) Canis Minor. It has a white dwarf (p. 75) companion. Also known as **Alpha Canis Minoris**.

Proxima Centauri the closest to us of the three stars that make up Alpha Centauri (p. 92). Proxima Centauri is an 11th-magnitude (p. 68) red dwarf (p. 72) that lies about one-tenth of a light year (p. 91) closer to us than the other two stars of Alpha Centauri, thereby making it the closest star of all to the Sun.

Rigel (*n*) the seventh-brightest star in the sky, of magnitude (p. 68) 0.12. Rigel is a blue-white supergiant (p. 72) lying 910 light years (p. 91) away in the constellation (↓) of Orion. Also known as **Beta Orionis**.

Sirius (*n*) the brightest star in the sky, of magnitude (p. 68) −1.46 in the constellation (↓) Canis Major. It is a white star that lies 8.7 light years (p. 91) away, which makes it the fifth-closest star to the Sun. It has a white dwarf (p. 75) companion. Also known as **Alpha Canis Majoris**.

Spica (*n*) a blue-white star of magnitude (p. 68) 1.0 in the constellation (↓) Virgo. Spica lies 260 light years (p. 91) away. Also known as **Alpha Virginis**.

Vega[1] (*n*) the fifth-brightest star in the sky, of magnitude (p. 68) 0.03. Vega is a blue-white star lying 26 light years (p. 91) away in the constellation (↓) Lyra. Also known as **Alpha Lyrae**.

Wolf 359 the third-closest star to the Sun, 7.6 light years (p. 91) away. It is a 13th-magnitude (p. 68) red dwarf (p. 72) in the constellation (↓) Leo.

the nearest stars

name	constellation	distance (light years)
Alpha Centauri	Centaurus	4.3
Barnard's star	Ophiuchus	6.0
Wolf 359	Leo	7.6
Lalande 21185	Ursa Major	8.1
Sirius	Canis Major	8.7
UV Ceti	Cetus	8.9
Ross 154	Sagittarius	9.5
Ross 248	Andromeda	10.3
Epsilon Eridani	Eridanus	10.7
Luyten 789-6	Aquarius	10.8
Ross 128	Virgo	10.8
61 Cygni	Cygnus	11.1
Epsilon Indi	Indus	11.2
Procyon	Canis Minor	11.3

constellation (*n*) any one of the areas into which the sky is divided for the purposes of finding and naming objects. A total of 88 constellations fills the whole sky. The first constellations were named by people thousands of years ago after their gods and heroes. Other constellations have been added more recently, named after scientific instruments. The boundaries of each constellation are fixed by the International Astronomical Union, astronomy's governing body. Each constellation has a Latin name. For a complete list of constellations, *see* appendix p. 198.

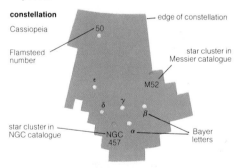

asterism (*n*) a group of stars, smaller than a constellation (↑), that forms a recognizable pattern, e.g. the shape of the Plough or Big Dipper is an asterism that forms part of the constellation of Ursa Major.

96 · STARS/CHARTS & CATALOGUES

zodiac (*n*) the 12 constellations (p. 95) through which the Sun passes each year, i.e. Aries, Taurus, Gemini, Cancer, Leo, Virgo, Libra, Scorpius, Sagittarius, Capricornus, Aquarius, Pisces. **zodiacal** (*adj*).

star catalogue a book that lists the positions, brightness and other information about stars.

star atlas a collection of maps of the sky, showing stars and constellations (p. 95).

Bayer letter a letter of the Greek alphabet that is given to each of the brightest stars in a constellation (p. 95), e.g. Betelgeuse is also known as Alpha Orionis, meaning alpha of Orion. The system was started in 1603 by the German astronomer Johann Bayer.

Flamsteed number a number given to each of nearly 3,000 stars in a catalogue published in 1725 by the English astronomer John Flamsteed, e.g. the Flamsteed number for Betelgeuse is 58 Orionis. The stars in each constellation (p. 95) are numbered in order of right ascension (p. 8).

fundamental star a star whose position and proper motion (p. 89) are known exactly, so that the positions of other stars can be measured with respect to it. There are about 1,500 of them scattered over the whole celestial sphere.

fundamental catalogue a star catalogue that contains a list of fundamental stars (↑).

FK a series of fundamental catalogues (↑) produced in Germany and used by astronomers the world over. The latest one is the FK5.

AGK a series of star catalogues published by the Astronomischen Gesellschaft in Germany, containing information on nearly 200,000 stars.

Bonner Durchmusterung BD. A star catalogue produced in Bonn, Germany, listing over 300,000 stars in the northern half of the sky.

Cordoba Durchmusterung CD. A star catalogue produced in Cordoba, Argentina, that lists over 600,000 stars in the southern hemisphere. The Cordoba Durchmusterung and the Bonner Durchmusterung (↑) together cover the entire sky.

Boss General Catalogue GC. A star catalogue listing over 30,000 stars, prepared by the American astronomers Lewis and Benjamin Boss.

SAO catalogue a catalogue listing over 250,000 stars published by the Smithsonian Astrophysical Observatory in Massachusetts.

Palomar Sky Survey a photographic atlas of the northern hemisphere of the sky and part of the southern hemisphere, made at the Palomar Observatory, California.

Southern Sky Survey a photographic atlas of the complete southern hemisphere of the sky, made jointly by the European Southern Observatory in Chile and the United Kingdom Schmidt telescope (p. 130) at Siding Spring, Australia.

Henry Draper catalogue HD. A catalogue that lists the spectral type (p. 73) of stars, made at Harvard College Observatory, Massachusetts.

ADS a catalogue of over 17,000 double stars (p. 85) produced by the American astronomer R.G. Aitken.

Astronomical Almanac a book published yearly containing tables giving the positions of the Sun, Moon and planets, as well as predictions of eclipses and other events.

almanac (*n*) a calendar (p. 52) giving dates and times of coming events in the sky.

Messier catalogue a list of over 100 star clusters (p. 88), nebulae and galaxies made by the French astronomer Charles Messier. Objects are known by their number in Messier's catalogue, e.g. the Crab nebula (p. 77) is M 1, and the Andromeda galaxy (p. 105) is M 31.

NGC New General Catalogue of Nebulae and Clusters of Stars. This lists nearly 8,000 objects.

IC Index Catalogue. Two catalogues that list added objects to those in the NGC (↑).

nebula (*n*) a cloud of gas and dust in space. Nebulae consist mostly of hydrogen, the most plentiful substance in the Universe, with about 10 per cent of helium and very small amounts of other substances. **nebulae** (*pl*).

bright nebula a nebula that is lit up by the bright stars that lie in or near it. *See also* emission nebula (↓), reflection nebula (↓).

dark nebula a nebula in which grains of dust block light from stars and other objects behind, e.g. the Coalsack (↓).

absorption nebula = dark nebula (↑).

emission nebula a nebula that gives out light of its own. Atoms (p. 116) of gas in the nebula are ionized (p. 116) by the energy of nearby stars. These ionized atoms give off light, which makes the nebula shine. HII regions (↓), planetary nebulae (p. 100) and supernova remnants (p. 77) are all emission nebluae.

HI region a nebula consisting of cold hydrogen gas, with a temperature of about 100 kelvins. HI regions do not give out light, but they do give out radio waves (p. 120) with a wavelength of 21 cm, known as the hydrogen line, which can be detected by radio telescopes. The gas in an HI region is much more thinly spread than in an HII region (↓), and it is not ionized (p. 116).

HII region an emission nebula (↑) consisting of hot, ionized (p. 116) hydrogen gas with a temperature of about 10,000 kelvins. The atoms (p. 116) of gas are ionized by ultraviolet (p. 120) light from stars inside the nebula.

dark nebula

Horsehead nebula, a dark nebula in constellation of Orion

HI and HII regions

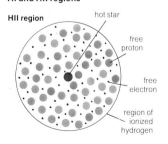

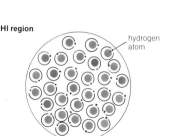

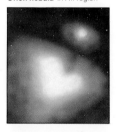

Orion nebula in HII region

Tarantula nebula

Strömgren sphere an HII region (↑) around a hot, bright star in a nebula. Strömgren spheres are roughly rounded in shape. Their size depends on the density of the gas and the temperature of the star.

reflection nebula a nebula that reflects the light of stars nearby. The light is reflected from dust grains in the nebula, so that the nebula appears bright.

diffuse nebula a misty-looking nebula with an irregular shape, e.g. the Orion nebula (↓). Stars are born inside diffuse nebulae.

Orion nebula the brightest and most famous HII region (↑) in the sky, visible as a misty patch in the constellation (p. 95) Orion. The Orion nebula lies about 1,500 light years (p. 91) away, and is about 15 light years in diameter. Stars are being born inside the nebula.

Trapezium (*n*) a group of four stars at the centre of the Orion nebula (↑), born recently from the nebula's gas. The brightest and hottest of the four stars lights up the Orion nebula.

Tarantula nebula a large and bright HII region (↑) in the Large Magellanic Cloud (p. 102), shaped like a spider.

Herbig-Haro object a small, bright nebula that changes in size, shape and brightness over a number of years. Herbig-Haro objects are thought to contain young stars in the process of formation. Herbig-Haro objects are similar to T Tauri stars (p. 84), but younger.

globule (*n*) a small, round dark nebula (↑). Globules are thought to be stars in the process of formation.

Bok globule = globule (↑).

Coalsack (*n*) a large dark nebula (↑) in the constellation (p. 95) of Crux, the southern cross. The Coalsack is the most noticeable of all the dark nebulae, for it blocks out light from a large number of stars in the Milky Way behind. It lies about 400 light years (p. 91) away and is 40 light years across.

interstellar medium thinly spread gas and dust in the space between the stars of the Galaxy, e.g. HI regions (↑).

100 · NEBULAE

interstellar absorption dark absorption lines (p. 121) in the spectra (p. 119) of light from distant stars, caused by gas between the stars.

interstellar extinction the dimming of light from distant stars caused by dust in the Galaxy. The dust particles scatter (p. 46) starlight, making it fainter and redder.

interstellar molecule a molecule (p. 116) that is found in nebulae. The molecules give out radio waves (p. 120) that can be picked up by radio telescopes (p. 134). Over 50 molecules are known to exist in space, many of them containing carbon.

interstellar grain a particle of dust in a nebulae. The dust grains cause interstellar extinction (↑). Interstellar grains are very small, about one ten-thousandth of a millimetre across. They are thought to consist largely of carbon.

planetary nebula a shell of gas thrown off from a star. The rounded shape of the nebula looks like the outline of a planet, which is how planetary nebulae got their name; they are actually nothing to do with planets. Planetary nebulae are emission nebulae (p. 98), lit up by a hot star at their centre. A planetary nebula is thought to be formed by gas thrown off from a red giant (p. 72) at the end of its life. The hot central part of the red giant remains behind, lighting up the nebula. Over many thousands of years the gas of the nebula moves off into space. The central star slowly cools down to become a white dwarf (p. 75).

Ring nebula a famous planetary nebula (↑) in the constellation (p. 95) of Lyra, looking like a smoke ring in space.

Gum nebula an emission nebula (p. 98) in the southern constellations (p. 95) of Puppis and Vela, named after the Australian astronomer Colin Gum who discovered it. The Gum nebula is thought to be a supernova remnant (p. 77).

Veil nebula an emission nebula (p. 98) in the constellation (p. 95) Cygnus formed by the explosion of a supernova (p. 76) about 60,000 years ago. It is part of the 'Cygnus loop'.

spiral nebula an old name for a galaxy.

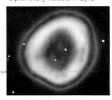

Ring nebula
a planetary nebula in Lyra

Veil nebula in Cygnus

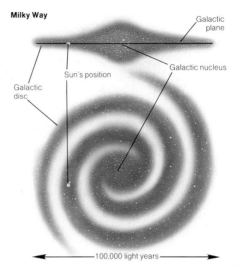

Milky Way (1) the common name for our Galaxy. The Milky Way Galaxy contains about 100,000 million stars and measures 100,000 light years (p. 91) in diameter; (2) the faint band of light seen stretching across the heavens, caused by all the distant stars in the galactic disc (↓).

galactic disc the flat plane in which most of the stars in our Galaxy lie, i.e. the path of the Milky Way.

galactic plane the central plane of the galactic disc (↑). *See also* galactic equator (p. 102).

galactic nucleus the rounded central part of our Galaxy, containing old stars of Population II (p. 75).

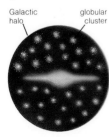

galactic halo a rounded volume of space around our Galaxy containing a scattering of old stars, thin gas and globular clusters (p. 88).

disc population a group of relatively young stars found in the galactic disc (↑), i.e. stars of Population I (p. 75).

halo population a group of old stars found in the galactic halo (↑), i.e. stars of Population II (p. 75).

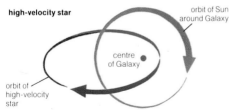

high-velocity star a star of the halo population (p. 101) that has a speed greater than 65 km s^{-1} relative to the local standard of rest (p. 90). This is because it is moving on its own highly elliptical (p. 16) orbit around the centre of the Galaxy, and does not share the galactic rotation (↓) of the Sun and its neighbouring stars.

galactic centre the centre of the Galaxy which, as seen from Earth, lies in the direction of the constellation (p. 95) of Sagittarius.

Sagittarius A a strong radio source (p. 134) in the constellation (p. 95) Sagittarius that marks the centre of our Galaxy.

galactic rotation the rotation of the Galaxy about its centre. The Galaxy's rotation is differential (p. 26), so that objects closest to its centre orbit most quickly, and those farthest out move the slowest. The Sun orbits the Galaxy with a speed of about 250 km s^{-1}, taking about 200 million years to complete one orbit.

galactic year the time taken for the Sun to orbit once around the Galaxy, i.e. roughly 200 million Earth years.

galactic coordinates a system of coordinates (p. 8) for measuring the position of objects in the Galaxy as seen from Earth. The coordinates used are galactic latitude (↓) and galactic longitude (↓).

galactic equator a great circle (p. 10) marking the plane of the Galaxy. The galactic equator is inclined (p. 19) at an angle of 62.6° to the celestial equator (p. 8).

galactic latitude the angle in degrees north or south of the galactic equator (↑), from 0° at the galactic equator to 90° at the galactic poles (↓); symbol b.

spiral arms of the Galaxy

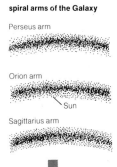

Perseus arm

Orion arm

Sun

Sagittarius arm

centre of Galaxy

galactic longitude the angle in degrees measured eastwards around the galactic equator (↑), from the direction of the galactic centre (↑); symbol *l*.

galactic pole either of the two points that lie 90° from the galactic equator (↑). The north galactic pole lies in the constellation (p. 95) of Coma Berenices; the south galactic pole lies in Sculptor.

Orion arm the spiral arm (p. 104) of our Galaxy in which the Sun lies, about 30,000 light years (p. 91) from the Galaxy's centre.

Perseus arm the spiral arm (p. 104) of our Galaxy that lies about 6,000 light years (p. 91) further from the Galaxy's centre than the Orion arm (↑).

Sagittarius arm the spiral arm (p. 104) of our Galaxy that lies about 6,000 light years (p. 91) nearer to the Galaxy's centre than the Orion arm (↑).

Gould's belt a band of hot, bright stars at an angle of about 15° to the galactic equator (↑). Gould's belt is an area of star birth in the Orion arm (↑) of our Galaxy.

zone of avoidance an area either side of the galactic plane (p. 101) in which no distant galaxies are seen, because dust in the Milky Way blocks the light from them.

extragalactic (*adj*) outside our Galaxy.

Magellanic Cloud either of two small galaxies next to the Milky Way. They are the nearest galaxies to us.

Large Magellanic Cloud an irregular-shaped galaxy 160,000 light years (p. 91) away in the constellation (p. 95) of Dorado. The Large Magellanic Cloud contains about 10,000 million stars, one-tenth of the number in our own Galaxy, and is about 35,000 light years in diameter.

Small Magellanic Cloud an irregular-shaped galaxy 190,000 light years (p. 91) away in the constellation (p. 95) Tucana. The Small Magellanic Cloud contains about 1,000 million stars, and is about 20,000 light years in diameter.

Magellanic Stream a large cloud of cold hydrogen gas that lies around both of the Magellanic Clouds (↑).

galaxy (*n*) a system of millions or billions of stars, as well as nebulae of dust and gas, held together by gravity (p. 112). Galaxies are of three main shapes: spiral galaxies (↓), elliptical galaxies (↓), and irregular galaxies (↓). Our own Galaxy, also known as the Milky Way, is written with a capital G. **galactic** (*adj*).

spiral galaxy a galaxy with arms of stars and gas that curve out from its centre. Our own Galaxy is a spiral galaxy.

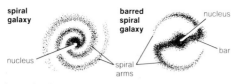

barred spiral galaxy a form of spiral galaxy (↑) with a straight bar of stars and gas across its centre. The spiral arms (↓) start at the ends of the bar. About a third of all spiral galaxies are barred spirals.

spiral arm a long, curving stream of stars and nebulae that gives spiral galaxies their shape.

density wave a kind of wave that is thought to move around a galaxy, giving rise to the galaxy's spiral arms (↑). The density wave presses together the gas in the spiral arms, causing stars to be formed.

elliptical galaxy a galaxy without spiral arms (↑). Elliptical galaxies are rounded in shape like a ball or an egg. They consist mostly of old stars, with very little gas.

supergiant elliptical galaxy the largest form of galaxy known in the Universe, containing a million million stars or more; they are also known as cD galaxies. Supergiant ellipticals are usually found at the centre of clusters of galaxies (p. 106).

cD galaxy = supergiant elliptical galaxy (↑).

irregular galaxy a galaxy with no particular shape. Irregular galaxies are usually small, like the Magellanic Clouds (p. 103).

lenticular galaxy a galaxy that is midway between spiral (↑) and elliptical (↑) in shape.

Hubble classification a way of arranging and naming galaxies according to their shape, i.e. spiral galaxies (↑) are given the letter S, barred spirals (↑) are classified as SB, elliptical galaxies (↑) are given the letter E, and irregular galaxies (↑) are called Irr. Spiral and barred spiral galaxies are given the letters a, b, and c depending whether their spiral arms (↑) are tight, medium, or loosely curved around the galaxy. Elliptical galaxies are numbered from E0 to E7, from the roundest ones to the most flattened ones. Lenticular galaxies (↑) are classified as S0 or SB0.

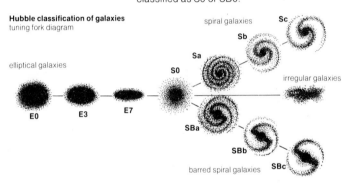

Hubble classification of galaxies
tuning fork diagram

tuning-fork diagram a Y-shaped drawing showing the various kinds of galaxies in the Hubble classification (↑).

Andromeda galaxy a large and bright spiral galaxy (↑) about 2.2 million light (p. 91) years away, the nearest large galaxy to us. It can be seen as a misty patch in the constellation (p. 95) Andromeda, and is the most distant object visible to the naked eye. The Andromeda galaxy is thought to contain about twice as many stars as our own Milky Way Galaxy.

Andromeda galaxy

GALAXIES/CLUSTERS OF GALAXIES

Local Group a group of about 30 galaxies held together by gravity (p. 112). The Andromeda galaxy (p. 105) and our own Galaxy are the largest members of the group.

cluster of galaxies a group of galaxies, containing from a few dozen to a few thousand members. Most galaxies are members of clusters. The Local Group (↑) is an example of a small cluster. The largest clusters are usually centered on a supergiant elliptical galaxy (p. 104). Clusters of galaxies are held together by gravity (p. 112).

virial theorem a way of calculating the total mass of a cluster of galaxies (↑) from the speed of movement of each of its member galaxies. The results show that clusters contain about 10 times as much mass as we can actually see in the form of galaxies. *See also* missing mass (↓).

missing mass the difference between the mass of a cluster of galaxies (↑) as calculated by the virial theorem (↑) and as expected from observations of the galaxies. The missing mass might be in the form of black holes (p. 79), unseen gas or faint stars.

Virgo cluster the nearest large cluster of galaxies (↑) to us. It lies 65 million light years (p. 91) away in the constellation (p. 95) of Virgo and contains about 2,500 known members.

Coma cluster a large cluster of galaxies (↑) containing at least 1,000 bright members, lying 400 million light years (p. 91) away in the constellation (p. 95) of Coma Berenices.

supercluster (*n*) a group of clusters of galaxies (↑). Superclusters measure several hundred million light years (p. 91) from side to side.

local supercluster the supercluster (↑) containing the Local Group (↑) and the Virgo cluster (↑) as well as over 50 other clusters of galaxies (↑).

Seyfert galaxy a galaxy with a very bright centre containing hot gas moving outwards at high speeds. The centres of Seyfert galaxies vary in brightness every few months. Most Seyfert galaxies are spiral galaxies (p. 104).

N galaxy a distant galaxy with a small, very bright centre. N galaxies are thought to be elliptical (p. 104), and are closely related to quasars (↓).

radio galaxy Cygnus A

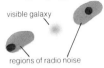

visible galaxy

regions of radio noise

Markarian galaxy a galaxy that gives out unusually large amounts of ultraviolet (p. 120) light. Many Markarian galaxies are also Seyfert galaxies (↑).

radio galaxy a galaxy that gives out large amounts of radio waves (p. 120). Most of the radio noise comes from large clouds of gas either side of the galaxy that appear to have been thrown out by explosions. Radio galaxies are supergiant elliptical galaxies (p. 104).

Cygnus A the first radio galaxy (↑) to be found, and the strongest. It lies in the constellation (p. 95) Cygnus at a distance of about 500 million light years (p. 91)!

radio galaxy Centaurus A

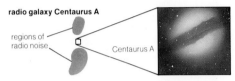

regions of radio noise

Centaurus A

Centaurus A the second-strongest radio galaxy (↑), also known as the supergiant elliptical galaxy (p. 104) NGC 5128. It lies 15 million light years (p. 91) away in the constellation (p. 95) Centaurus and is the nearest radio galaxy to us. It has a band of dark dust crossing it.

quasar a star-like object with a very high redshift (p. 115). Quasars are thought to be powerful forms of Seyfert galaxies (↑) and N galaxies (↑) so distant that usually only their bright centres can be seen. Quasars give out as much energy as hundreds of normal galaxies from a volume of space no bigger than the solar system. Some quasars give out radio waves (p. 120), but most do not.

quasi-stellar object QSO = quasar (↑).

quasi-stellar radio source QSS. A quasar (↑) that gives out radio waves (p. 120).

BL Lacertae object an object related to quasars (↑), but closer to us. It looks like a star, changes rapidly in brightness, and gives out radio waves (p. 120). It is thought to be an elliptical galaxy (p. 104) with a very bright centre.

Lacertid = BL Lacertae object (↑).

THE UNIVERSE/EXPANSION

Universe (*n*) all matter, space and time. Everything that we can see is part of the Universe.

cosmology (*n*) the study of the Universe, its beginning, and how it has changed with time.

cosmological (*adj*) of the Universe.

cosmogony (*n*) the study of the birth and growth of particular objects in the Universe, such as galaxies, stars and the solar system.

expansion of the Universe the fact that the whole Universe seems to be getting bigger. This is because the galaxies outside the Local Group (p. 106) all appear to be moving away from each other, as shown by the redshift (p. 115) in their light.

expansion of the Universe

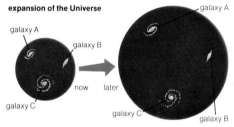

cosmological principle the idea that we are in no special place in the Universe. Therefore to someone living in another galaxy far away, the Universe would look much the same as it does to us.

Hubble's law the law that governs the expansion of the Universe (1). According to Hubble's law, the more distant a galaxy is from us, the faster it is moving away.

Hubble's constant a figure that relates the speed of a galaxy's movement away from us to its distance in the Universe; symbol H_0. Hubble's constant is difficult to measure, and so is not known exactly. According to present-day figures, Hubble's constant is between 50 and 100 kilometres per second per million parsecs (p. 91). In other words, galaxies move away from us at a speed of between 50 and 100 km s^{-1} for each million parsecs they are distant.

Hubble's law

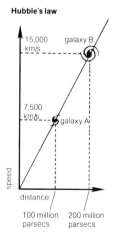

Hubble time the age of the Universe if Hubble's constant (↑) has remained the same since the Big Bang (p. 110). The Hubble time is $1/H_0$, which gives an age for the Universe of between 10 and 20 billion years with present-day figures for Hubble's constant. In practice, the expansion of the Universe (↑) was probably faster in the past, so the Universe is somewhat younger than the Hubble time.

deceleration parameter a quantity that describes the slowing down of the expansion of the Universe (↑); symbol q_0. The best present-day measurements show that there has been little slowing down in the expansion of the Universe since the Big Bang (p. 110).

lookback time the time taken for light to reach us from distant objects in the Universe. When we look far away into space at objects billions of light years (p. 91) away, we are in effect looking back billions of years in time.

Olbers' paradox a question put forward in 1826 by the German astronomer Heinrich Olbers. He asked: If the Universe is endless then the number of stars in it should also be endless, so why does not the light from all these stars add up to make the night sky bright? The answer to Olbers' paradox is that the Universe does not contain enough stars and galaxies to cover the whole sky, and that they do not live long enough to fill the Universe with light. It is sometimes said that the expansion of the Universe (↑) is the answer to Olbers' paradox but this is not so.

open Universe the condition in which the expansion of the Universe (↑) will continue for ever. Present-day measurements of the deceleration parameter (↑) show that the Universe is open.

closed Universe the condition in which the expansion of the Universe (↑) will stop at some time in the future, and the Universe will then start to get smaller again. Present-day measurements of the deceleration parameter (↑) show that the expansion of the Universe is not slowing down enough for it ever to stop. *See* open Universe (↑).

110 · THE UNIVERSE/EVOLUTION

Big Bang theory

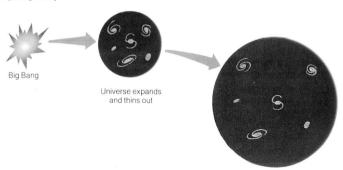

Big Bang

Universe expands and thins out

Big Bang the supposed event that started the expansion of the Universe (p. 108). According to the Big Bang theory, the Universe as we know it began with a very large explosion between 10 billion and 20 billion years ago, and the Universe has been getting bigger ever since.

primeval fireball the hot, dense ball which contained all the matter in the Universe at the time of the Big Bang (↑).

cosmic background radiation electromagnetic radiation (p. 118) from space, picked up at radio and infrared wavelengths (p. 119). The radiation is like that given out by a body with a temperature of 2.7 degrees above absolute zero (p. 118). Therefore the Universe is not completely cold, but has a temperature of 2.7 K. The cosmic background radiation is the result of heat left over from the Big Bang (↑), given out at the time of decoupling (↓).

isotropic (*adj*) of something that is the same in all directions, e.g. the cosmic background radiation (↑) is isotropic. **isotropy** (*n*).

radiation era the time from a few seconds to about 30,000 years after the Big Bang (↑) when most of the energy in the Universe was in the form of high-energy photons (p. 118).

matter era the time in which most of the energy in the Universe has been in the form of matter. The matter era started about 30,000 years after the Big Bang (↑).

oscillating Universe

decoupling (*n*) the time when radiation in the Universe was no longer absorbed (p. 119) by matter. The decoupling of radiation and matter happened about 300,000 years after the Big Bang (↑), when the density of the Universe had dropped to a certain point as a result of the expansion of the Universe (p. 108). **decouple** (*v*).

hadron era the first few millionths of a second after the Big Bang (↑) when heavy atomic (p. 116) particles such as protons (p. 116) and neutrons (p. 116) were formed.

lepton era the first second or so after the hadron era (↑) in which light atomic (p. 116) particles including electrons (p. 116) were formed.

oscillating Universe the state in which the Universe would get bigger and smaller in turn, going from one Big Bang (↑) to another. This could only happen in a closed Universe (p. 109).

steady state theory the idea that the Universe never had a beginning, nor will it have an end. The steady state theory says that the Universe has always existed and that it has always looked much like it does today. According to this theory, new matter comes into being all the time to fill the space left by the expansion of the Universe (p. 108). This theory is not widely accepted. *See also* continuous creation (↓).

steady state theory

Universe gets bigger but does not thin out

continuous creation the production of new matter out of nothing, as supposed by the steady state theory (↑). On this theory, new matter comes into being all the time in the form of hydrogen atoms (p. 116) from which new galaxies grow, so that the density of the Universe remains the same at all times.

astrophysics (*n*) the study of the physical nature of stars, galaxies and the Universe. **astrophysical** (*adj*).

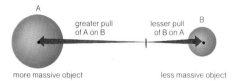

gravity (*n*) a force that tries to pull objects towards each other. The more massive an object, the stronger the pull of its gravity. Gravity acts as though all the mass of a body is in one point at the body's centre. **gravitational** (*adj*).

gravitation (*n*) = gravity (↑).

gravitational field the volume of space around a body in which the pull of that body's gravity (↑) can be felt. The force of gravity gets less with distance from the body according to the inverse-square law (p. 114).

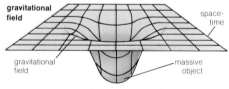

gravitational collapse the event when a body gets smaller as a result of the inward pull of its own gravity (↑), e.g. a cloud of gas undergoes gravitational collapse to become a star, or a star undergoes gravitational collapse at the end of its life to become a white dwarf (p. 75), a neutron star (p. 76) or a black hole (p. 79).

gravitational wave a form of energy given out by massive objects undergoing acceleration, in a similar way to the radiation that is given out by charged particles being accelerated, e.g. by large stars exploding as supernovae (p. 76) or objects falling into black holes (p. 79). The existence of gravitational waves is predicted by the general theory of relativity (↓).

relativity (*n*) the name given to two theories to do with space, time and matter put forward by Albert Einstein: the special theory of relativity (↓) in 1905, and the general theory of relativity (↓) in 1916.

relativistic (*adj*) of something moving close to the speed of light, e.g. the particles that make up cosmic rays (p. 117) are moving at relativistic velocities.

special theory of relativity a theory that relates measurement of quantities such as time, length and mass made by one observer to measurements of the same quantities by a different observer who is moving at constant velocity with respect to the first observer. The special theory of relativity says that mass can be changed into energy, and the other way round, as given by the equation $E = mc^2$, where E is energy, m is mass and c is the speed of light. According to the special theory of relativity, things cannot travel faster than the speed of light.

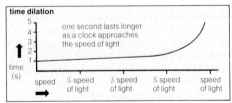

time dilation the slowing down of time for objects that are moving close to the speed of light. Time dilation is predicted by the special theory of relativity (↑).

general theory of relativity a theory that deals with matter and gravity (↑). According to the general theory of relativity, gravity is actually a bending or curving of space-time (↓) caused by the presence of matter.

space-time a system of coordinates (p. 8) that describes the position (p. 8) of an object in time as well as space. According to the theory of relativity (↑), space and time are not separate, but must always be dealt with together.

world line the path of an object in space-time (p. 113).

gravitational lens the bending of a beam of light in a gravitational field (p. 112).

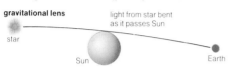

inverse-square law a law which states that the strength of a force, e.g. gravity (p. 112), or a flow of energy, e.g. light, gets weaker with the square of the distance from the source: if an object is moved twice as far away, the strength of the light or gravity received from it falls by four times; if its distance were three times greater, the strength of its light or gravity would be nine times less, and so on.

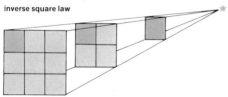

world line of the Earth's orbit around the Sun

Doppler effect the change in wavelength (p. 119) of light or other forms of electromagnetic radiation (p. 118) given out by a moving object. If the object is moving away, its light is stretched out to longer wavelengths, i.e. a redshift (↓). If the object is coming closer, its light gets shorter in wavelength, i.e. a blueshift (↓).

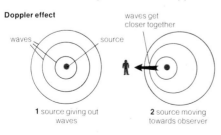
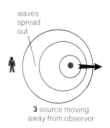

ASTROPHYSICS/DOPPLER EFFECT

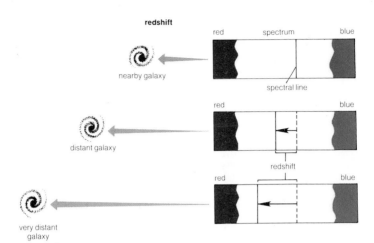

redshift a lengthening of the wavelength (p. 119) of light or other forms of electromagnetic radiation (p. 118) from a source that is moving away. The effect is known as a redshift because a lengthening of wavelengths moves the light towards the red end of the spectrum (p. 119).

blueshift a shortening of the wavelength (p. 119) of light or other forms of electromagnetic radiation (p. 118) received from an object that is moving towards the observer. The effect is known as a blueshift because the light is moved towards the blue (shorter-wavelength) end of the spectrum (p. 119).

gravitational redshift a lengthening of the wavelength (p. 119) of light or other forms of electromagnetic radiation (p. 118) caused by the gravitational field (p. 112) around the source.

ion (*n*) an atom (p. 116) that has lost or gained one or more electrons (p. 116). **ionized** (*adj*).

positive ion an atom (p. 116) that has lost electrons (p. 116), and so has a positive electric charge.

negative ion an atom (p. 116) that has gained electrons (p. 116), and so has a negative electric charge.

ionization (*n*) the act of removing or adding electrons (↓) to an atom (↓) to make ions (p. 115). In astronomy, ionization happens as a result of high temperatures, e.g. in stars, or when high-energy particles hit atoms. **Ionize** (*v*), **ionized** (*adj*).

photoionization (*n*) the ionization (↑) of atoms (↓) by high-energy photons (p. 118), e.g. in an HII region (p. 98).

plasma (*n*) a gas made of ionized (↑) atoms (↓) and free electrons (↓). Most of the matter in the Universe is in the form of a plasma.

recombination (*n*) the event when electrons (↓) join up with positive ions (p. 115). It is the opposite of ionization (↑).

neutral atom an atom (↓) that has not been ionized (↑).

atom (*n*) the smallest part of a chemical element (↓) that has the characteristics of that element. An atom consists of a central nucleus (↓) with electrons (↓) around it.

element¹ (*n*) a substance consisting of atoms (↑) which each have the same number of protons (↓) in their nucleus (↓). Over 100 different elements are known.

nucleus¹ (*n*) the central part of an atom (↑), consisting of protons (↓) and neutrons (↓). The nucleus is the heaviest part of the atom. **nuclei** (*pl*), **nuclear** (*adj*).

proton (*n*) a particle that carries a positive electric charge. Protons are found in the nucleus (↑) of an atom (↑).

neutron (*n*) a particle that carries no electric charge. Neutrons are found in the nucleus (↑) of an atom (↑). The mass of a neutron is almost the same as that of a proton (↑).

electron (*n*) a particle with a negative electric charge. The amount of charge is equal to that of a proton (↑). Electrons move around the nucleus (↑) of an atom (↑). The mass of an electron is 1/1836 that of a proton or neutron (↑).

molecule (*n*) a grouping of two or more atoms (↑) joined together.

heavy element any element (↑) heavier than hydrogen or helium.

ionization

hydrogen atom
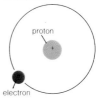

molecule

hydrogen molecule

water molecule

 = hydrogen

 = oxygen

cosmic abundance of the elements

element	number of atoms relative to 1,000,000 atoms of hydrogen
hydrogen	1,000,000
helium	63,000
oxygen	690
carbon	420
nitrogen	87
silicon	45
magnesium	40
neon	37
iron	32
sulphur	16

cosmic abundance of elements the relative number of atoms (↑) of each element (↑) in the Universe. *See* table.

nucleosynthesis (*n*) the formation of elements (↑) by nuclear reactions (p. 70). Nucleosynthesis of helium took place in the Big Bang (p. 110). Nucleosynthesis of the heavy elements (↑) happens inside stars.

cosmic rays atomic (↑) particles moving through space near to the speed of light. Most are protons (↑), but the nuclei (↑) of all elements (↑) are present in smaller numbers among cosmic rays, as are electrons (↑). Most are thought to come from supernovae (p. 76), although some come from flares (p. 31) on the Sun.

bremsstrahlung (*n*) electromagnetic radiation (p. 118), usually at radio wavelengths (p. 119), given out by fast-moving electrons (↑) as they are slowed down or change direction. This happens as they pass near protons (↑) in an ionized (↑) gas.

bremsstrahlung

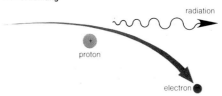

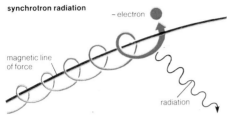

synchrotron radiation electromagnetic radiation (↓) given out by fast-moving electrons (p. 116) as they move around in a strong magnetic field. Synchrotron radiation is the cause of the radio noise given out by most radio sources (↓) in the Universe.

Compton effect the increase in wavelength (↓) of a photon (↓) when it hits an electron (p. 116) and loses some of its energy. When photons hit relativistic (p. 113) electrons they gain energy and their wavelength gets shorter; this is known as the *inverse Compton effect*, and it is thought to be how X-rays (p. 120) are produced in supernovae (p. 76) and quasars (p. 107).

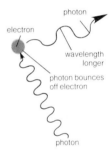

temperature (*n*) a measure of the amount of heat that something contains.

absolute zero the coldest possible temperature, −273°C; 0 K.

Kelvin scale a scale of temperature that starts at absolute zero (↑); symbol K. Kelvins are the same as Celsius degrees.

electromagnetic radiation a flow of energy produced when electrically charged particles change speed or direction. It includes radio waves (p. 120), infrared radiation (p. 120), visible light (p. 120), ultraviolet radiation (p. 120), X-rays (p. 120) and gamma rays (p. 120). The only difference between all these forms of electromagnetic radiation is their wavelength (↓) and frequency (↓). All electromagnetic radiation travels at the speed of light. Electromagnetic radiation has the characteristics both of a wave and a particle.

photon (*n*) a particle of electromagnetic radiation (↑).

ASTROPHYSICS/THE SPECTRUM · 119

spectrum of electromagnetic radiation

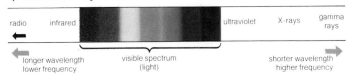

source

wavelength

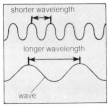

spectrum (n) (1) the complete range of wavelengths (↓) of electromagnetic radiation (↑), from radio waves (p. 120) to gamma rays (p. 120); (2) the coloured band produced when light from a star or other object is passed through a spectroscope (p. 132). **spectra** (pl), **spectral** (adj).

source (n) something that gives out electromagnetic radiation (↑) or any other form of energy, e.g. the Sun is a source of light and heat.

emit (v) to send or give out something, e.g. light or atomic (p. 116) particles. **emission** (n).

radiate (v) to send or give out energy, e.g. electromagnetic radiation (↑) or atomic (p. 116) particles. **radiation** (n).

absorb (v) to take in energy, e.g. electromagnetic radiation (↑) or atomic (p. 116) particles. **absorption** (n).

wavelength (n) the distance from one point on a wave to the same point on the next wave; symbol λ. Wavelength is equal to the speed of the wave divided by its frequency (↓).

ångström (n) a unit of measure of the wavelength (↑) of light; symbol Å. One ångström = 10^{-10} metres. 10,000 ångströms = 1 micron (1μm).

frequency (n) the number of waves that pass a fixed point in a given time, usually one second. Frequency is equal to the speed of the waves divided by their wavelength (↑). It is measured in hertz (p. 120).

frequency

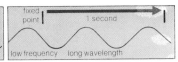

hertz (*n*) a unit of measure of the frequency (p. 119) of electromagnetic radiation (p. 118); symbol Hz. 1 hertz = one wave per second.

light (*n*) the part of the range of electromagnetic radiation (p. 118) that we can see with our eyes, sometimes also including infrared and ultraviolet radiation (↓). We see light of different wavelengths (p. 119) as different colours. *Visible light* has a wavelength between about 4,000 Å at the blue end of the spectrum (p. 119) to 7,000 Å at the red end of the spectrum.

velocity of light the speed of light in a vacuum is 299,792.5 km s^{-1}. All electromagnetic radiation (p. 119) travels at this speed.

radio waves electromagnetic radiation (p. 118) with wavelengths (p. 119) longer than about 1 mm.

infrared radiation electromagnetic radiation (p. 118) with wavelengths (p. 119) between about 1 mm and 7,000 Å.

ultraviolet radiation electromagnetic radiation (p. 118) with wavelengths (p. 119) between about 4,000 Å and 100 Å.

X-rays electromagnetic radiation (p. 118) with wavelengths (p. 119) between about 100 Å and 0.1 Å.

gamma rays electromagnetic radiation (p. 118) with wavelengths (p. 119) shorter than about 0.1 Å.

interference (*n*) the effect that happens when two sets of electromagnetic waves come together. When the waves arrive in step they add up to become stronger; when they arrive out of step they become weaker. All forms of electromagnetic radiation (p. 118) undergo interference.

interference fringes bright and dark bands of light caused by interference (↑).

diffraction (*n*) the slight bending of light around the edge of an object. Longer wavelengths (p. 119) are diffracted more than shorter ones.

spectroscopy (*n*) the study of light, e.g. from stars and galaxies, by means of a spectroscope (p. 132). Spectroscopy allows astronomers to find out what objects are made of (*see* spectral lines ↓), or how they are moving (*see* Doppler effect p. 114). **spectroscopic** (*adj*).

interference

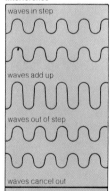

diffraction

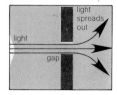

ASTROPHYSICS/THE SPECTRUM · 121

spectrogram (*n*) a photograph of a spectrum (p. 119).
continuous spectrum a spectrum (p. 119) that consists of a continuous range of wavelengths (p. 119), like a rainbow, rather than separate emission lines (↓).
continuum (*n*) = continuous spectrum (↑).
spectral lines narrow lines seen in an object's spectrum (p. 119). Each kind of atom (p. 116) has its own set of spectral lines, so that by studying the lines in an object's spectrum astronomers can tell what the object is made of. The lines can be either dark, i.e. absorption lines (↓), or bright, i.e. emission lines (↓).

continuous spectrum

emission lines

absorption lines

absorption lines dark lines crossing a continuous spectrum (↑), caused when light from a hot object passes through cooler gas, e.g. when light from a star passes through its own cooler outer regions. Each line is the result of a wavelength (p. 119) of light that has been absorbed (p. 119) by atoms (p. 116) of the cooler gas.
Fraunhofer lines absorption lines (↑) in the spectrum (p. 119) of the Sun.
emission lines particular wavelengths (p. 119) of electromagnetic radiation (p. 118) given out by atoms (p. 116) of a gas. When seen through a spectroscope (p. 132), emission lines can either appear as bright lines on a continuous spectrum (↑), e.g. when given out by hotter gas around a star, or they can be seen on their own, e.g. when given out by a nebula.
emission spectrum a spectrum (p. 119) consisting of bright emission lines (↑), e.g. given out by the gas in a nebula.

forbidden lines emission lines (p. 121) given out by a gas that has a much lower density than is possible on Earth. The low densities needed are found only in space, e.g. in a nebula, so under conditions on Earth the emission of the lines is said to be forbidden.

Zeeman effect the dividing of spectral lines (p. 121) into two or more parts by a magnetic field.

polarized (*adj*) of electromagnetic radiation (p. 118) whose waves all lie in one plane. Light can be polarized by magnetic fields, e.g. synchrotron radiation (p. 118) is polarized, or by being reflected, e.g. off dust grains in nebulae. **polarize** (*v*), **polarization** (*n*).

black body an object that is both a perfect emitter (p. 119) and a perfect absorber (p. 119) of radiation, i.e. it absorbs all the radiation that falls on it without reflecting any, and it emits radiation in the way predicted by Planck's law (↓). Stars behave like black bodies.

Zeeman effect

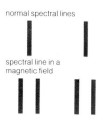

normal spectral lines

spectral line in a magnetic field

Planck's law

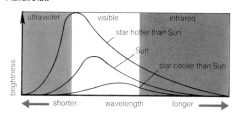

Planck's law a law that describes how much electromagnetic radiation (p. 118) is given out at each wavelength (p. 119) by a gas of a certain temperature. The hotter the gas, the more radiation that is given out at each wavelength. The wavelength at which most radiation is given out gets shorter with increasing temperature, i.e. the coolest objects emit (p. 119) most in the infrared (p. 120) while the hottest objects emit most in the ultraviolet (p. 120).

effective temperature the surface temperature that a star would have if it behaved exactly like a black body (↑) with the same size and the same energy output as itself.

ASTRONOMICAL INSTRUMENTS/OPTICS · 123

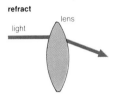
refract

concave

convex

reflect

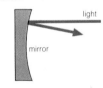
focal length

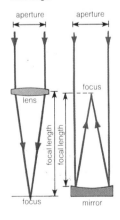

optics (*n*) (1) the study of light and vision; (2) the lenses and mirrors used in instruments, e.g. the optics of a telescope. **optical** (*adj*).

lens (*n*) a piece of glass with curved surfaces that refracts (↓) light.

refract (*v*) to bend the path of a beam of light, e.g with a lens. **refraction** (*n*), **refracting** (*adj*).

mirror (*n*) a smooth piece of glass with a flat or curved surface that reflects (↓) light. A mirror is given a thin coating of silver or aluminium to make it reflective.

speculum (*n*) an old word for the mirror of a telescope, used in the days when mirrors were made of shiny metal.

concave (*adj*) of a surface, e.g. of a lens or mirror, that curves inwards.

convex (*adj*) of a surface, e.g. of a lens or mirror, that curves outwards.

flat (*n*) a mirror that has a flat surface.

reflect (*v*) to throw back a beam of electro-magnetic radiation (p. 118) from a surface, e.g. a mirror reflects light. **reflection** (*n*), **reflecting** (*adj*), **reflective** (*adj*).

objective (*n*) the lens at the front of a refracting telescope (p. 127) that collects and focuses (↓) light.

object glass = objective (↑).

aperture (*n*) the width of the objective (↑) or main mirror in a telescope. The larger the aperture of a telescope, the fainter the objects that can be seen and the more detail that can be seen.

eyepiece (*n*) a lens or group of lenses through which a person looks to see the image (↓) in a telescope. An eyepiece magnifies (p. 124) the image.

image (*n*) the picture of an object formed by a lens or a mirror.

focus^2 (*n*) the point at which light beams come together to form a sharp image (↑) after having been refracted (↑) or reflected (↑). **foci** (*pl*), **focal** (*adj*), **focus** (*v*).

focal length the distance between a mirror or a lens and the point at which it brings parallel lines of light to a focus (↑). Lines of light from a very distant object are parallel.

focal plane the flat surface on which a telescope forms an image (p. 123). Some instruments, e.g. the Schmidt telescope (p. 130), form an image on a curved surface, known as the *focal surface*.

focal ratio the focal length (p. 123) of a lens or mirror divided by its aperture (p. 123), e.g. a lens with a focal length of 60 cm and an aperture of 10 cm has a focal ratio of 6, written as f.6. The bigger the focal ratio the larger the image (p. 123) but the smaller the field of view (↓). Also known as **f ratio**.

field of view the width, in degrees, of the scene that is visible through a telescope.

prime focus the point at which the main mirror in a reflecting telescope (p. 127), or the objective (p. 123) in a refracting telescope (p. 127), brings light to a focus (p. 123).

coudé focus

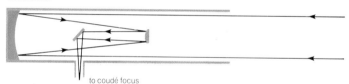

to coudé focus

coudé focus a fixed point, outside the tube of a reflecting telescope (p. 127), to which light collected by the telescope is reflected by mirrors. The coudé focus is a good place for putting large and heavy equipment, such as spectroscopes (p. 132), that cannot be moved.

magnification

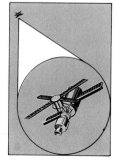

magnification (*n*) the amount by which an object seems larger when seen through a telescope, e.g. a telescope that makes something appear ten times bigger is said to have a magnification of ten times (written 10×). Magnification is calculated by dividing the focal length (p. 123) of the telescope by the focal length of the eyepiece (p. 123), e.g. a telescope of focal length 60 cm and an eyepiece of focal length 1 cm gives a magnification of 60×. An eyepiece of different focal length on the same telescope would give a different magnification. **magnify** (*v*).

ASTRONOMICAL INSTRUMENTS/ABERRATION · 125

resolution

two close stars, images resolved

two close stars images not resolved

resolution (*n*) the ability of a telescope to see fine detail, or to separate two things, e.g. stars, that appear close together. The resolution of a telescope is usually measured in seconds of arc (p. 15). **resolve** (*v*).

resolving power = resolution (↑).

Dawes limit a way of calculating the resolution (↑) of a telescope. According to the Dawes limit, the resolution of a telescope in seconds of arc (p. 15) is 11.5 divided by the aperture (p. 123) of the telescope in centimetres, e.g. a telescope of 10 cm aperture has a resolution of 1.15 arc seconds.

aberration (*n*) a fault or imperfection in a lens or mirror that affects the quality of the image (p. 123) it produces.

chromatic aberration

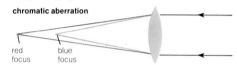

red focus   blue focus

chromatic aberration coloured edges around an object seen through a lens, resulting from the fact that a lens does not bring light of all colours to the same focus (p. 123). Light of shorter (blue) wavelengths (p. 119) is bent more than light of longer (red) wavelengths, so that the blue light comes to a focus closer to the lens than does the red light. The effect of chromatic aberration can be lessened by putting together two lenses, each made of glass with a different composition. Mirrors do not suffer from chromatic aberration.

achromatic (*adj*) of a lens made of two pieces of glass to stop chromatic aberration (↑).

doublet

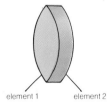

element 1   element 2

doublet (*n*) a lens that is made up of two pieces of glass, e.g. an achromatic (↑) lens is a doublet.

element^2 (*n*) one of the pieces of glass that makes up a lens, e.g. a doublet (↑) is a two-element lens.

apochromat (*n*) a lens made of three or more elements (↑) to prevent chromatic aberration (↑) at all wavelengths (p. 119).

spherical aberration

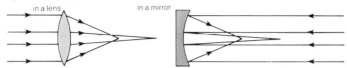

spherical aberration the failure of different parts of a lens or mirror to bring light to the same focus (p. 123), resulting in an image (p. 123) that is not sharp. The central part of a lens or a mirror with a spherical curve has a longer focal length (p. 123) than the outer parts. Spherical aberration can be prevented in a mirror by making the centre deeper, so that the curve on the mirror becomes parabolic (p. 19). In lenses, spherical aberration is avoided by putting two different lenses together.

coma[2] (*n*) an aberration (p. 125) in which star images (p. 123) are spread out into pear shapes towards the edge of the field of view.

aplanatic (*adj*) of a lens that has been corrected for chromatic aberration (p. 125), spherical aberration (↑) and coma (↑).

astigmatism (*n*) an aberration (p. 125) in which light from a star is focused (p. 123) not into a point but into lines, so that the star appears cross-shaped. **astigmatic** (*adj*).

collimation (*n*) the act of lining up the lenses and mirrors in an optical instrument to give the best images (p. 123). **collimate** (*v*).

diffraction ring any of a number of circles of light around the image (p. 123) of a star seen through a telescope, resulting from diffraction (p. 120) in the telescope. At the centre of the diffraction rings is the Airy disc (↓).

Airy disc the rounded image (p. 123) of a star seen through a telescope. The star does not appear as a point because of diffraction (p. 120) in the telescope. The size of the Airy disc depends on the aperture (p. 123) of the telescope, becoming smaller as the aperture gets larger. The smaller the Airy disc, the better the telescope's resolution (p. 125). Around the Airy disc are seen diffraction rings (↑).

diffraction rings

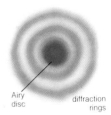

ASTRONOMICAL INSTRUMENTS/TELESCOPES · 127

diffraction grating a series of thin lines closely spaced on a piece of glass that produces a spectrum (p. 119) by diffraction (p. 120).

dispersion (*n*) (1) the spreading out of light into a spectrum (p. 119) by a prism (↓) or a diffraction grating (↑); (2) a measure of how widely the spectrum is spread out, i.e. *high dispersion, low dispersion*.

prism

prism (*n*) a triangular shaped block of glass that can spread light out into a spectrum (p. 119) or that can reflect (p. 123) light.

objective prism a thin prism (↑) which is placed in front of a telescope lens or mirror to produce a spectrum (p. 119) of each star in the field of view (p. 124).

telescope (*n*) an instrument for looking at objects far away. Telescopes use lenses or mirrors to collect and focus (p. 123) light. Telescopes collect more light than the human eye, so they can show fainter objects. They also make objects look larger, so that small details can be seen.

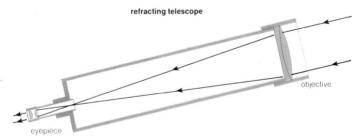

refracting telescope

objective

eyepiece

refracting telescope a telescope in which a lens at the front collects and focuses (p. 123) light.
refractor (*n*) = refracting telescope (↑).
reflecting telescope a telescope in which a concave (p. 123) mirror collects and focuses (p. 123) light. The largest telescopes are of the reflecting (p. 123) kind, because big mirrors are easier and cheaper to make than big lenses.
reflector (*n*) (1) = reflecting telescope (↑); (2) something that reflects (p. 123).

secondary mirror a smaller mirror onto which light is reflected (p. 123) by the main mirror in a reflecting telescope (p. 127). The secondary mirror can have a flat surface, as in a Newtonian reflector (↓), or a curved surface, as in a Cassegrain reflector (↓). The purpose of a secondary mirror is to reflect light into an eyepiece outside the telescope tube, and to fold a long focal length (p. 123) into a relatively short tube.

paraboloid

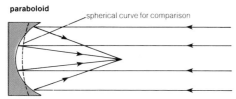
spherical curve for comparison

paraboloid (*n*) a mirror whose surface is curved like a parabola (p. 19). A paraboloid is free from spherical aberration (p. 126). The main mirror is a paraboloid in most reflecting telescopes (p. 127), e.g. the Newtonian reflector (↓) and the Cassegrain reflector (↓).

hyperboloid (*n*) a mirror with a surface curved like a hyperbola (p. 19). The secondary mirror (↑) in a Cassegrain reflector (↓) is a hyperboloid.

Newtonian reflector a reflecting telescope (p. 127) in which the light collected by the main mirror is reflected (p. 123) through a hole in the side of the tube by a flat secondary mirror (↑) or by a prism (p. 127). Most small reflecting telescopes are of the Newtonian kind.

Newtonian reflector

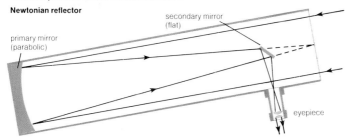

Cassegrain reflector

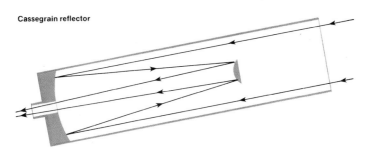

Cassegrain reflector a reflecting telescope (p. 123) in which the secondary mirror (↑) has a convex (p. 123) surface and reflects (p. 123) light back through a hole in the middle of the main mirror; this point is known as the Cassegrain focus. In a Cassegrain reflector, the main mirror is a paraboloid (↑) and the secondary mirror is a hyperboloid (↑). Most large modern telescopes use the Cassegrain system.

Gregorian reflector an old form of reflecting telescope (p. 123), similar to the Cassegrain (↑), in which the secondary mirror (↑) is concave (p. 123) in shape.

Ritchey-Chrétien reflector a Cassegrain reflector (↑) in which both the main and secondary mirrors (↑) are hyperboloids (↑). This system gives a wide field of view (p. 124) free of coma (p. 126).

Dall-Kirkham reflector a Cassegrain reflector (↑) in which the secondary mirror (↑) has a curve that is spherical and the main mirror has a curve that is between a sphere and a paraboloid (↑). The system is widely used, since the mirrors are easier to make than in a regular Cassegrain.

catadioptric telescope a telescope that uses both lenses and mirrors to form an image (p. 123), e.g. the Schmidt telescope (p. 130) and the Maksutov telescope (p. 130).

corrector plate a thin lens used to correct the aberrations (p. 125) of a telescope mirror and to give it a wider field of view (p. 124), as in a catadioptric telescope (↑).

Schmidt telescope

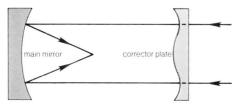

Schmidt telescope a catadioptric telescope
(p. 129) used for taking wide-angle photographs
of the sky. The Schmidt telescope has a main
mirror with a spherical curve. To correct for
spherical aberration (p. 126) and to provide a
wide field of view (p. 124) free of coma (p. 126),
a corrector plate (p. 129) is placed at the front
of the telescope tube.

Schmidt-Cassegrain a Schmidt telescope (↑)
with a secondary mirror (p. 128) like that in a
Cassegrain (p. 129) telescope.

Maksutov telescope a Schmidt telescope (↑) in
which the corrector plate (p. 129) is deeply
curved. Often the centre of this lens is given a
reflective (p. 123) spot so that the telescope
acts like a Cassegrain (p. 129).

astrograph (n) a telescope that is used for taking
pictures of the sky, especially a refracting
telescope (p. 127) used for astrometry (p. 89).

coelostat (n) an instrument in which a flat mirror
reflects (p. 123) light from the sky into a fixed
telescope. The flat mirror is driven around a
polar axis (p. 132) so that it follows an object
across the sky as the Earth rotates. Usually the
light is reflected off a second flat mirror before
it enters the telescope. Coelostats are often
used for observing the Sun.

heliostat (n) an instrument similar to the coelostat
(↑) in which the first flat mirror can be tilted to
view different declinations (p. 8).

siderostat (n) a heliostat (↑) in which light from
the first mirror is reflected (p. 123) along a line
parallel to the Earth's axis, either directly into a
fixed telescope or off another mirror into a fixed
telescope.

Maksutov-Cassegrain

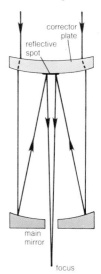

coronagraph (*n*) a refracting telescope (p. 127) used to observe the Sun's corona (p. 32). Inside the telescope, a rounded object blocks out the bright light from the Sun's surface to produce the same effect as at an eclipse.

heliometer (*n*) a refracting telescope (p. 127), no longer used, in which the objective (p. 123) is divided into two, each half producing a separate image (p. 123). The two halves of the objective can be moved to bring the two images together, and the amount of movement needed shows the size of an object or how far two objects are apart. It was used for measuring the diameter of the Sun and the angle between stars.

mounting (*n*) the support that holds a telescope and allows it to be pointed at various parts of the sky.

altazimuth mounting a simple telescope mounting (↑) in which the telescope can point freely up and down, i.e. in altitude (p. 11), and swing from side to side, i.e. in azimuth (p. 11). The telescope must be constantly moved both in altitude and in azimuth to track an object as it crosses the sky.

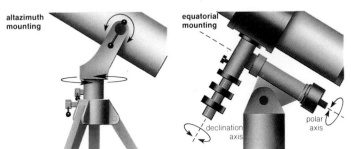

equatorial mounting a telescope mounting (↑) in which one axis, known as the polar axis (p. 132), points to the celestial pole (p. 9). The polar axis is turned so that the telescope follows an object across the sky as the Earth spins. The second axis, known as the declination axis (p. 132), does not have to be moved once an object is in the telescope's view.

polar axis the axis in an equatorial mounting (p. 131) that points to the celestial pole (p. 9), i.e. it is lined up parallel to the Earth's axis.

declination axis the axis in an equatorial mounting (p. 131) that allows the telescope to point in declination (p. 8). The declination axis is at right angles to the polar axis (↑).

setting circle a marking on the polar axis (↑) and declination axis (↑) of an equatorially mounted (p. 131) telescope that allows the telescope to aim at an object with known coordinates (p. 8).

spectroscope (*n*) an instrument in which light is spread out into a spectrum (p. 119) by a prism (p. 127), or a diffraction grating (p. 127). The spectroscope known as a spectrograph (↓) is usually used. **spectroscopic** (*adj*).

spectroscope

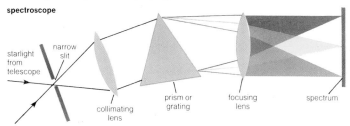

spectrograph (*n*) a spectroscope (↑) in which the spectrum (p. 119) is recorded on a photograph. Spectrographs are widely used in astronomy. They are usually placed at the Cassegrain (p. 129) or coudé focus (p. 124) of a telescope.

spectroheliograph (*n*) an instrument for photographing the Sun at one particular wavelength (p. 119) of light.

spectrometer (*n*) (1) an instrument that measures the positions of various spectral lines (p. 121); (2) an instrument that records a spectrum (p. 119) electronically.

spectrophotometer (*n*) an instrument for measuring the brightness of a spectrum (p. 119) at each wavelength (p. 119).

photometer (*n*) an instrument for measuring the amount of light received from an object, e.g. the brightness of a star.

ASTRONOMICAL INSTRUMENTS/PHOTOELECTRICITY · 133

photoelectric cell

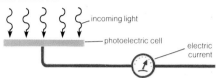

- **photoelectric cell** a device that turns light into an electric current. The photoelectric cell gives out electrons (p. 116) when light hits it. The more light that hits the cell, the greater the number of electrons that are given out.
- **photocell** (*n*) = photoelectric cell (↑).
- **photomultiplier** (*n*) a device that amplifies (electronically increases) the electric current given out by a photoelectric cell (↑). This is necessary in astronomy because the light received from stars etc is so faint that it produces only a very small electric current.
- **bolometer** (*n*) an instrument that measures the total amount of electromagnetic radiation (p. 118) coming from an object at all wavelengths (p. 119).
- **image intensifier** a device like a television camera that electronically increases the brightness of an image (p. 123) seen through a telescope.
- **charge-coupled device** CCD. A silicon chip that electronically records an image (p. 123) of the light that hits it. CCD are over ten times as sensitive as photographic plates.
- **interferometer** (*n*) an instrument, or pair of instruments, in which two beams of light or other radiation from an object are brought together to cause interference (p. 120). An interferometer can resolve (p. 125) details too small to be seen by one telescope alone. Interferometers are used to measure small angles such as the diameters of stars and the distance apart of two close stars.
- **speckle interferometry** a way of adding together a series of photographs of the same object taken with one telescope to produce a resolution (p. 125) as good as if the Earth's atmosphere were not in the way.

comparator

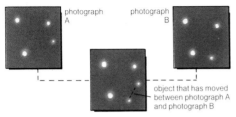

photograph A
photograph B
object that has moved between photograph A and photograph B

comparator (n) a device for examining two photographs of the same area of sky to look for any changes, e.g. the movement of an asteroid or the difference in brightness of a variable star (p. 80).

blink microscope a comparator (↑) in which first one and then the other photograph are brought into view through an eyepiece (p. 123) in quick succession. Any objects that have moved between the photographs will appear to jump backwards and forwards.

planetarium (n) (1) an instrument that throws a picture of the night sky onto the inside of a dome; (2) the building in which such an instrument is housed.

radio astronomy the study of radio waves (p. 120) given out naturally by objects in space. Radio waves with wavelengths (p. 119) between about 1 cm and 30 metres pass through the Earth's atmosphere. These radio waves are collected by radio telescopes (↓).

radio source an object that gives out radio waves (p. 120). Supernova remnants (p. 77), pulsars (p. 78), quasars (p. 107), and certain galaxies are strong radio sources.

radio telescope an instrument for collecting radio waves (p. 120) coming from objects in space. A radio telescope consists of an antenna (↓) and an amplifier (↓). Radio telescopes are much larger than optical telescopes because the wavelength (p. 119) of radio waves is much longer than that of light waves. To improve their resolution (p. 125), a pair of radio telescopes are often joined together to make radio interferometers (↓).

radio telescope

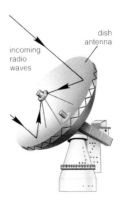

dish antenna
incoming radio waves

antenna (*n*) the part of a radio telescope (↑) that collects incoming radio waves (p. 120). The antenna can be a metal dish, shaped so that it acts like the mirror of a reflecting telescope (p. 127), or it can be a simple metal bar, known as a *dipole*; in practice, many dipoles are joined together over a large area to increase the sensitivity of the radio telescope. **antennae** (*pl*).

aerial (*n*) = antenna (↑).

array (*n*) a number of radio antennae (↑) joined together to make a large radio telescope (↑).

amplifier (*n*) a device that increases the strength of an electric current, e.g. as produced by the radio waves (p. 120) received by a radio telescope (↑). **amplify** (*v*).

beamwidth (*n*) a way of measuring the resolution (p. 125) of a radio telescope (↑): the smaller the beamwidth, the better the resolution. The beamwidth of a radio telescope is governed by its aperture (p. 123) and the wavelength (p. 119) of the radio waves (p. 120) it is receiving.

microwaves (*n.pl.*) radio waves (p. 120) with wavelengths (p. 119) between about 1 mm and 1 m, as observed by radio telescopes (↑).

radio interferometer a pair of radio telescopes (↑) joined together to give better resolution (p. 125) than is possible with each telescope on its own. The radio waves (p. 120) collected by each telescope are brought together to cause interference (p. 120) that tells astronomers about the object's shape.

Mills Cross a radio interferometer (↑) consisting of two antennae (↑) arranged in the shape of a cross.

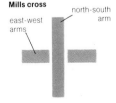

Mills cross
east-west arms
north-south arm

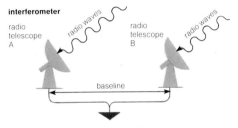

interferometer
radio telescope A
radio waves
radio telescope B
radio waves
baseline

136 · ASTRONOMICAL INSTRUMENTS/RADIO ASTRONOMY

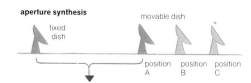

aperture synthesis

aperture synthesis a way of building up the view of the sky that would be seen with one very large radio telescope (p. 134). Several smaller telescopes are joined together in pairs to form radio interferometers (p. 135), and one or more of the telescopes is moved around to change the length of the baseline (↓). The aperture (p. 123) of the imaginary larger telescope is thereby filled in step by step. The observations are added together in a computer to make a complete map of the object under study. The resolution (p. 125) of such a system is the same as that of a single radio telescope whose diameter equals that of the longest baseline used, e.g. several kilometres.

baseline (*n*) the distance between the two telescopes in a radio interferometer (p. 135).

very-long baseline interferometry VLBI. The practice of collecting observations of the same object made at the same time by two radio telescopes (p. 134) a great distance apart, e.g. on different continents. The observations are recorded on magnetic tape and are later added together. Very-long baseline interferometry makes possible a radio interferometer (p. 135) with a diameter as great as that of the Earth, showing objects in far greater detail than can be obtained in any other way.

very-long baseline interferometry

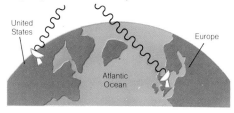

ASTRONOMICAL INSTRUMENTS/OBSERVATORIES · 137

molecular line a particular wavelength (p. 119) at which interstellar molecules (p. 116) give out radio waves (p. 120).

hydrogen line the wavelength (p. 119) of 21 cm at which non-ionized (p. 116) hydrogen gives out radio waves (p. 120).

radar astronomy the study of objects in the solar system, e.g. planets and their moons, by reflecting (p. 123) radio waves (p. 120) off them. A strong burst of radio waves is sent out from a radio telescope (p. 134) on Earth. The time taken for the reflected beam to return tells how far away the object is. Radar can also be used to map the surface of planets and to measure their speed of rotation.

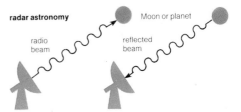

radar astronomy

observatory (*n*) a place from which studies of the sky are made.

dome² (*n*) a building in which a telescope is kept. A dome has a rounded roof that can be opened to allow the telescope to view the sky.

observatory

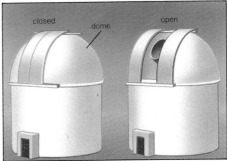

geocentric system

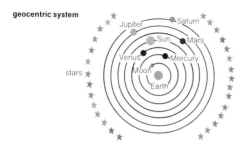

geocentric system the old belief that the Earth was the centre of the Universe. The Earth was thought to stand still while everything else went around it each day. This idea was generally accepted until the 17th century.

heliocentric system the theory that the Sun is the centre of the Universe. This idea took over from the geocentric system (1) in the 17th century, when it was realized that the Earth is an ordinary planet going around the Sun. Later it was found that not even the Sun is the centre of the Universe – in fact, the Universe has no centre.

heliocentric system

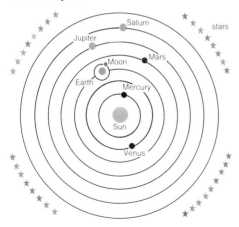

HISTORY OF ASTRONOMY · 139

Ptolemaic system a form of geocentric system (↑) put forward in the second century AD by Ptolemy. According to Ptolemy, the Sun, Moon and planets moved around the Earth along a series of circles, known as deferents (↓) and epicycles (↓). By putting these together in the right way, the Ptolemaic system could account for the observed motions of the planets. It was widely accepted for over 1,300 years.

Tychonic system a theory of the Universe put forward around 1580 by the Dane, Tycho Brahe. In his view, the Sun went around the Earth, but the other planets went around the Sun.

Copernican system a heliocentric system (↑) put forward in 1543 by the Polish astronomer Nicolaus Copernicus. According to Copernicus, the Sun was the centre of the Universe, and the Earth went around it like any other planet. However, Copernicus still kept to the old idea that the planets moved along circular paths. Kepler's laws (p. 17) finally showed that the paths of the planets are ellipses (p. 16).

deferent (*n*) in the Ptolemaic system (↑), the circular path around the Earth that was followed by the centre of a smaller circle, the epicycle (↓). In some cases, the centre of the deferent was not at the centre of the Earth.

epicycle (*n*) in the Ptolemaic system (↑), a small circular motion made by a planet, while the centre of the epicycle moved along a larger circle called the deferent (↑).

archaeoastronomy (*n*) the study of the possible astronomical purposes of ancient buildings, such as Stonehenge (↓) and the pyramids of Egypt. Many rings of standing stones are thought to have been used as simple observatories to watch the rising and setting of the Sun and Moon.

Stonehenge (*n*) a circle of large stones on Salisbury Plain, England, put up over 3,000 years ago. An outer stone, called the Heel stone, is lined up towards the point where the Sun rises at the summer solstice (p. 10), so that Stonehenge may have been used as a kind of observatory for keeping a calendar (p. 52).

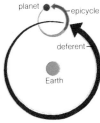

deferent and epicycle

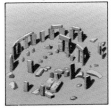

Stonehenge

armillary sphere an old astronomical instrument. It consists of a series of metal rings showing great circles (p. 10) on the sky, e.g. the celestial equator (p. 8), the ecliptic (p. 9) and the horizon. Once a star had been sighted, the astronomer could read off its right ascension (p. 8) and declination (p. 8) from the armillary sphere.

astrolabe (*n*) an old astronomical instrument consisting of two metal plates, one on top of the other. By turning one plate to show the altitude (p. 11) of a given star, the time of night could be read off.

quadrant (*n*) an old instrument in the shape of a quarter circle, with markings so that the altitude (p. 11) of a star could be read off. The astronomer sighted the star along a movable arm on the quadrant.

mural quadrant a quadrant (↑) fixed to a wall that is lined up in a north-south direction, i.e. in the observer's meridian (p. 10). The altitude (p. 11) of stars was observed as they crossed the meridian. The mural quadrant was an early form of transit instrument (p. 56).

astrology (*n*) the belief that events in the sky, e.g. the changing positions of the planets, can affect the lives of people on Earth. **astrological** (*adj*).

mural quadrant

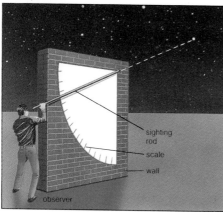

ROCKETS/PROPULSION · 141

rocket (*n*) (1) an engine that works by reaction propulsion (↓); (2) the device that contains such an engine. Rockets carry propellant (p. 143) which they burn to produce thrust (p. 145). Rockets are used to launch (p. 148) objects and to travel in space.

missile (*n*) a rocket that is used as a weapon, or that carries a weapon.

ballistic missile a missile (↑) that is under power only at launch (p. 148); it then follows a curving path to its target. It does not go into orbit.

ballistic (*adj*) of the trajectory (p. 16) followed by a rocket or missile (↑) once its engines have been turned off and it is acted upon only by the pull of gravity (p. 112).

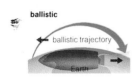
ballistic

launch vehicle any device that is used to put a satellite or space probe into space, i.e. a rocket that is not used as a weapon.

sounding rocket a small rocket that carries instruments to study the Earth's upper atmosphere, or to make astronomical observations from above the atmosphere. A sounding rocket goes up through the atmosphere, but unlike a launch vehicle (↑) does not go into orbit.

sounding rocket

propulsion (*n*) the act of pushing something forward. **propel** (*v*).

reaction propulsion a form of propulsion (↑) in which an object is pushed along by the force of a stream of gas escaping from it, as happens in a rocket. The action of the gas escaping in one direction produces a reaction that pushes the object in the opposite direction.

reaction propulsion

jet propulsion = reaction propulsion (↑). **jet propelled** (*adj*).

stage (*n*) a part of a rocket that has its own engines and propellants (p. 143). Rockets are usually made up of two or three stages, each of which is self-contained. The stages fire one after the other. Once one stage has used up its propellant it falls away and the stage above it then takes over.

multi-stage rocket a rocket which is made up of stages (↑).

step rocket = multi-stage rocket (↑).

142 · ROCKETS/PROPULSION

booster (*n*) (1) the large and powerful first stage (p. 141) of a multi-stage rocket (p. 141); (2) an extra rocket engine that fires for a short time to help launch (p. 148) a large rocket.

strap-on booster one of two or more rocket engines fastened to the side of a main rocket, e.g. the Solid Rocket Boosters (p. 170) of the Shuttle (p. 169). Strap-on boosters work for a short time to help launch (p. 148) the main rocket. They fall away as the main rocket rises.

sustainer (*n*) an engine that keeps a rocket going after the boosters (↑) have fallen away.

mass ratio the ratio between the mass of a rocket when fully fuelled and the mass of the propellant (↓) that it contains.

pogo (*n*) the end-to-end shaking of a rocket that sometimes happens while the engine is firing.

payload (*n*) the mass of the satellite, space probe, scientific equipment, etc, that a launch vehicle (p. 141) can carry. The weight of the rocket itself or its propellant (↓) are not part of the payload.

fairing (*n*) the covering placed over the payload (↑) on top of a rocket to protect it during its flight through the atmosphere. It is jettisoned (p. 151) once the rocket is above the atmosphere.

shroud (*n*) = fairing (↑).

nose-cone (*n*) the conical part at the tip of a rocket. It may protect the payload (↑), like a fairing (↑), or it may form part of the payload itself. In the case of missiles (p. 141), the nose-cone is covered with a heat shield (p. 160) to withstand the heat of re-entry (p. 159) into the Earth's atmosphere.

backup (*n*) a second device or system that can take the place of another in case of a failure. Important parts of rockets and spacecraft have one or more backups. Even astronauts have backups who train for the same flight.

redundancy (*n*) the condition of having more than one way of doing something, so that if one piece of equipment fails another can take over. **redundant** (*adj*).

vernier rocket a small rocket used to help guide the course and speed of a larger rocket.

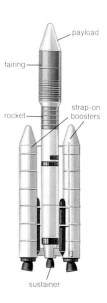

booster, payload and fairing

ROCKETS/ROCKET TYPES, PROPELLANT · 143

chemical rocket

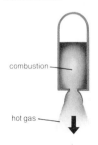

combustion

hot gas

electric rocket

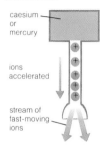

caesium or mercury

ions accelerated

stream of fast-moving ions

nuclear rocket

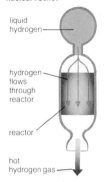

liquid hydrogen

hydrogen flows through reactor

reactor

hot hydrogen gas

chemical rocket a rocket that works by the combustion (p. 146) of a fuel. All the rockets that have been launched (p. 148) from Earth so far, and all those foreseen for use as launch vehicles (p. 141), are chemical rockets.

electric rocket a rocket whose exhaust (p. 147) consists of ionized (p. 116) particles. Atoms (p. 116) of a heavy element (p. 116) such as caesium or mercury are ionized by heat or electric fields. The ionized particles are then accelerated at high speed out of the rocket by electric fields. Electrical power for the rocket is provided by solar panels (p. 158). Electric rockets give a low thrust (p. 145) so they cannot be used for launches (p. 148) from Earth, but they are useful for moving about once in space and can reach much higher final velocities than chemical rockets (↑).

ion propulsion the form of propulsion (p. 141) used in an electric rocket (↑).

nuclear rocket a possible form of rocket for the future in which a nuclear reactor heats a gas to very high temperatures so that it escapes from the rocket at high speed. They might be used out in space, but not for take-off from Earth.

nuclear-electric rocket an electric rocket (↑) in which the electrical power is provided by a nuclear reactor.

nuclear pulse rocket a possible rocket of the future that would be pushed along by the force of small nuclear explosions set off behind it.

photon propulsion a possible way of using a beam of light, which consists of photons (p. 118), for propulsion (p. 141). The best form of photon propulsion would be provided by the radiation pressure (p. 61) of sunlight pushing on a large reflecting surface; this is known as solar sailing.

propellant (*n*) any substance used for the propulsion (p. 141) of a rocket. The propellant of a rocket consists of its fuel and an oxidizer (p. 144). Propellant can be liquid or solid.

fuel (*n*) a substance, liquid or solid, that is burned in a rocket to produce the hot gases that push the rocket along. *See also* liquid fuel (p. 144); solid fuel (p. 145). **fuel** (*v*).

oxidizer (*n*) a substance that allows the fuel in a rocket to burn in the airless conditions of space. An oxidizer usually, but not always, contains oxygen. In rockets that have liquid fuel (↓), the most common oxidizer is liquid oxygen. In solid-propellant rockets the oxidizer is mixed in with the fuel.

working fluid the substance that acts as the propellant (p. 143) in a nuclear rocket (p. 143) or an electric rocket (p. 143) where there is no combustion (p. 146). In such rockets, the working fluid is simply heated to form an exhaust (p. 147) of fast-moving gas.

bi-propellant (*n*) a form of liquid propellant (p. 143) in which the fuel and the oxidizer (↑) are kept separate. They are stored in different tanks and are forced into the combustion chamber (p. 146) by pumps.

mono-propellant (*n*) a form of rocket propellant (p. 143) in which the fuel and the oxidizer (↑) are mixed together. Solid propellants are mono-propellants; some liquid propellants are also mono-propellants.

liquid fuel

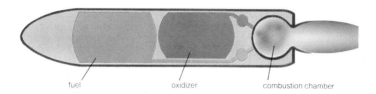

fuel   oxidizer   combustion chamber

liquid fuel a fuel in a rocket that is in liquid form. Common liquid fuels are kerosene and liquid hydrogen.

hydrazine (*n*) a rocket fuel containing hydrogen and nitrogen, chemical symbol N_2H_4.

cryogenic (*adj*) of rocket propellants (p. 143) that have to be kept at very low temperatures, e.g. liquid hydrogen, liquid oxygen. Cryogenic propellants are liquid only at very low temperatures.

solid fuel

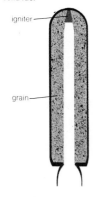

ullage

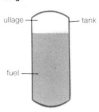

solid fuel a fuel in a rocket that is in the form of a solid. In practice, the solid fuel is mixed together with an oxidizer (↑) to form a solid propellant (p. 143). Rockets with solid propellants have fewer working parts than liquid-propellant rockets, and so are less likely to go wrong. But the burning of solid propellants cannot be controlled as easily as that of liquids, and they cannot be turned on and off at will.

grain (*n*) the solid propellant (p. 143) in a rocket. The grain is shaped to control the thrust (↓) of the rocket as the propellant burns.

tank (*n*) a container in a rocket in which liquid propellant (p. 143) is stored. The fuel and the oxidizer (↑) are kept in separate tanks.

ullage (*n*) the empty volume in a rocket's propellant (p. 143) tanks. Before a rocket engine is started in space, small rockets called ullage rockets are fired to make the propellants settle to the bottom of the partly empty tanks.

pump (*n*) a device that forces propellants (p. 143) into a rocket's combustion chamber (p. 146).

hypergolic (*adj*) of propellants (p. 143) that start to burn when they meet, without ignition (↓).

ignition (*n*) (1) the moment when the fuel in a rocket starts to burn; (2) the act of starting such burning, e.g. by an electrical spark or a small flame. **ignite** (*v*).

thrust (*n*) the force that pushes a rocket along. Thrust is produced by the hot gases escaping from a rocket engine. The amount of thrust depends on the mass of propellant (p. 143) being used in a given time and the exhaust velocity (p. 147). To be able to take off, a rocket must produce a thrust that is greater than its own weight.

specific impulse a measure of the amount of thrust (↑) produced by a given weight of rocket propellant (p. 143) each second. Specific impulse is measured in seconds, e.g. if a rocket engine produces 250 tonnes of thrust by burning 1 tonne of propellant per second, the rocket is said to have a specific impulse of 250 seconds. The higher the specific impulse, the better the performance of the rocket.

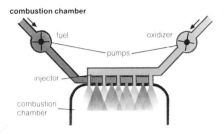

combustion (*n*) the act of burning something; in the case of a rocket, the burning of its fuel.

combustion chamber that part of a liquid-propellant rocket in which the fuel is burned. The fuel and the oxidizer (p. 144) are pumped into the combustion chamber where they are ignited (p. 145). The burning happens at high temperature and under high pressure, producing hot gases that provide the rocket's thrust (p. 145).

chamber pressure the pressure of the hot gases in a rocket's combustion chamber (↑). In the main engines of the Space Shuttle (p. 169) the chamber pressure is 200 times the Earth's atmospheric pressure, but other engines work at lower pressures than this.

injector (*n*) a device through which liquid propellants (p. 143) are forced into the combustion chamber (↑) of a rocket. An injector usually consists of a plate with holes in it through which the fuel and the oxidizer (p. 144) flow so that they are thoroughly mixed as they enter the combustion chamber.

mixture ratio the ratio between the weight of oxidizer (p. 144) and the weight of fuel pumped into the combustion chamber (↑) of a rocket. In the main engine of the Space Shuttle (p. 169), for example, the mixture ratio is 6 parts of liquid oxygen to 1 part of liquid hydrogen fuel.

throat (*n*) the narrow opening between the combustion chamber (↑) and the nozzle (↓) of a rocket engine.

expansion ratio the ratio of the area of the throat (↑) of the combustion chamber (↑) to the area of the nozzle (↓) at its widest.

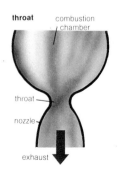

ROCKETS/ENGINES · **147**

area ratio = expansion ratio (↑).
exhaust (*n*) the stream of gas that flows out of a rocket engine, producing thrust (p. 145).
exhaust velocity the velocity of the gases escaping from a rocket nozzle (↓). The greater the exhaust velocity for given propellants (p. 143), the greater the thrust (p. 145).
nozzle (*n*) the bell-shaped opening at the rear of a rocket engine through which the exhaust (↑) gases escape, producing the thrust (p. 145) that pushes the rocket along. The nozzle is shaped to allow the escaping gases to expand and speed up as they pass through, producing an exhaust velocity (↑) greater than the speed of sound.
de Laval nozzle a form of rocket nozzle (↑) that first of all narrows down to a throat (↑), to increase the pressure of the escaping gases, then opens up again to allow the gases to escape at high speed. In practice, the throat is usually considered to be part of the combustion chamber (↑), and the word nozzle is used only for the wide opening at the rear.
convergent-divergent nozzle = de Laval nozzle (↑).
gimbal^1 (*v*) to tilt a rocket engine or its nozzle (↑) at a slight angle from side to side to guide the rocket on its course.
gimbal^2 (*n*) a device on which a rocket engine is set in place so that it can be gimballed.
thrust vector control the act of gimballing (↑) a rocket engine to control the direction of the exhaust (↑) and hence to guide the course of the rocket.
burn (*n*) the firing of a rocket engine.
burnout (*n*) the moment when a rocket has used up all its propellant (p. 143) and stops firing.
shutdown (*n*) the act of stopping the firing of a rocket engine.
cutoff (*n*) the moment when the flow of propellant (p. 143) to a rocket engine is turned off, after the engine has fired for a given time.
static firing the firing of a rocket engine on the ground to try it out. In a static firing the rocket engine is held down so that it does not take off.

gimbal

nozzle

← gimbal →

launch (v) to send a rocket and its payload (p. 142) away from the surface of the Earth, etc. **launch** (n).

launch window the length of time during which a rocket must be launched (↑) if it is to place a satellite into the right orbit or a space probe on the right trajectory (p. 16). Because the Earth is rotating, the launch pad (↓) is continually moving relative to the aiming point in space. The aiming point itself, such as the Moon or a planet, can also be moving, e.g. launch windows to Venus come around every 19 months and to Mars every 26 months. The launch window of the Space Shuttle (p. 169) is affected by the need for good light conditions at its landing areas.

launch pad the place from which a rocket leaves; often simply called the pad. The countdown (↓) takes place and liquid propellants (p. 143) are loaded into the rocket while it is standing on the launch pad.

launch pad

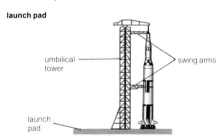

cosmodrome (n) the name given by the Soviet Union to place from which space launches (↑) are made.

umbilical tower the framework tower that stands next to a rocket on a launch pad (↑). Lines from the umbilical tower join the rocket to the ground before launch, carrying fuel and electrical supplies. Some umbilical towers are used to service the rocket while it is on the launch pad by means of swing arms (↓).

umbilical (adj) of a tube or pipe through which something flows, e.g. rocket propellants (p. 143), electrical power or air to breathe.

ROCKETS/LAUNCHING · 149

swing arm an arm from an umbilical tower (↑) that allows workers to reach various parts of a rocket while it is on the launch pad (↑), and along which astronauts enter their spacecraft; these arms swing out of the way before the rocket is launched (↑). Other arms carry fuel and electrical lines; these remain joined to the rocket until the last moment, and pull away at lift-off (↓).

countdown (*n*) the time leading up to the launch (↑) of a rocket. During the countdown a carefully planned series of events must be carried out in a set order to prepare the rocket (and its crew, if any) for launch. The countdown may begin one or two days before the rocket's launch. Time is counted backwards from the start of the countdown to the launch time, known as T = 0.

hold (*n*) a delay in a countdown (↑), e.g. to correct a fault or to wait for the weather to improve. Countdowns usually have built-in holds which allow the rocket to be launched (↑) on time even if things go wrong.

scrub (*v*) to put off a launch (↑) attempt to another day because of problems.

flame deflector an object placed at an angle under a rocket to turn the rocket's hot exhaust (p. 147) gases to one side.

hold-down arm a mechanical arm that holds a rocket down to the launch pad (↑) while its engines build up thrust (p. 145). The arms let go at the moment of lift-off (↓).

lift-off the moment when a rocket rises off its launch pad (↑).

ascent (*n*) the upward movement of a rocket into orbit. **ascend** (*v*).

lift-off

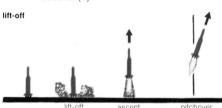

lift-off   ascent   pitchover

150 · ROCKETS/LAUNCHING

launch azimuth the compass heading that a rocket follows as it leaves the launch pad (p. 148).

pitchover (*n*) the moment shortly after launch (p. 148) when a rocket turns at an angle from upright so that it follows a curving path towards orbit.

abort (*v*) to cut short a space flight because of a failure, accident or other sudden danger, especially during the launch (p. 148). **abort** (*n*).

launch escape system a set of small rockets on a tower above a manned spacecraft. They fire to pull the spacecraft clear in case of a launch (p. 148) failure.

destruct (*n*) the act of intentionally blowing up or otherwise destroying a rocket that has gone off course, to prevent it from crashing on Earth and causing damage.

accelerate (*v*) to change the speed or direction of an object's movement.

acceleration (*n*) the increase in an object's velocity divided by the time taken to increase it, e.g. if an object accelerates from rest to 100 metres per second in 10 seconds it has undergone an acceleration of 100/10 = 10 metres per second for every second (10 m s^{-2}) it was accelerating.

g symbol for the acceleration due to the Earth's gravity, i.e. 9.8 metres per second every second (9.8 m s^{-2}); this is known as 1 *g*.

***g* force** the force felt by an astronaut during a space flight as a result of the acceleration or deceleration (p. 160) or the rocket. During launch on the Shuttle (p. 169) the astronaut suffers a force of up to 3 *g* (↑), i.e. the astronaut feels three times as heavy as on Earth; during re-entry (p. 159) on the Shuttle the force is no more than about 1.5 *g*. In orbit there is no feeling of gravity (p. 112); this is known as zero *g* (p. 152), and the astronaut feels weightless.

dynamic pressure the force of the Earth's atmosphere on a rocket as it accelerates upwards; symbol q.

max q the greatest dynamic pressure (↑) on a rocket during launch (p. 148), usually reached 70 to 80 seconds after lift-off (p. 149).

launch azimuth

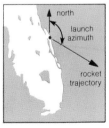

launch escape system
on an Apollo spacecraft

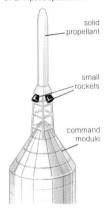

g force

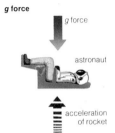

ROCKETS/LAUNCHING · 151

staging and separation

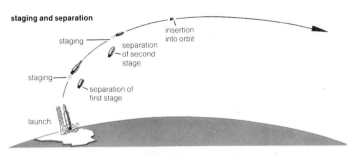

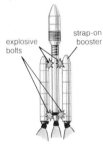

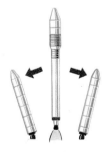

staging (*n*) the moment or act when one stage (p. 141) of a rocket stops burning and the next stage takes over.

separation (*n*) (1) the moment when a burned-out rocket stage (p. 141) falls away from the rest of the rocket; (2) the act of a stage falling away.

jettison (*v*) to throw aside or drop an unwanted object, e.g. an empty fuel tank or rocket stage (p. 141).

explosive bolt a fastening that explodes to jettison (↑) a rocket stage (p. 141).

downrange (*n*) the distance from a place of launch (p. 148) that a rocket has reached while on its way up through the atmosphere.

injection (*n*) the act of placing a spacecraft on a calculated trajectory (p. 16).

insertion (*n*) the act of placing a rocket or spacecraft into an orbit around the Earth or another body.

sub-orbital (*adj*) of the path of a rocket that does not go into orbit but which falls back to Earth, e.g. a ballistic (p. 141) trajectory (p. 16). The trajectory of a sounding rocket (p. 141) is sub-orbital.

apogee kick motor a small rocket engine that fires at a satellite's apogee (p. 18) to circularize the satellite's orbit at that height. Used especially to put satellites into geostationary orbit (p. 23).

deploy (*v*) to put something out for use, for example, to place a satellite into orbit from the payload bay (p. 171) of the Space Shuttle (p. 169), or to open up the solar panels (p. 158) of a satellite.

astronautics (n) the theory and practice of travelling in space.

spacecraft (n) any constructed device that flies in space, e.g. a rocket, satellite or space probe, whether with a crew or without.

boilerplate (n) a metal copy of a spacecraft, usually heavier than the real thing and containing few or no working parts. Boilerplate models are used for test purposes.

satellite[2] (n) = artificial satellite (↓).

artificial satellite a spacecraft put into orbit around the Earth. A space probe can also become an artificial satellite of the Moon or a planet if it goes into orbit around them.

space probe a spacecraft that is sent away from the Earth to study conditions in space, or to examine another body in space, e.g. the Moon or a planet.

weightlessness (n) the condition in which there is no feeling of weight. When an object is falling without anything to stop it, that object is weightless. A spacecraft in orbit is falling endlessly, either around the Earth or some other body such as the Sun, so that it and everything inside it is also weightless. Also known as **free fall** or **zero gravity**.

zero g the condition in which there is no g force (p. 150), i.e. weightlessness (↑).

free fall the condition in which an object is falling freely in a gravitational field (p. 112) with nothing to stop it, e.g. a satellite is in free fall around the Earth or a space probe is in free fall around the Sun. An object in free fall is weightless, *see* weightlessness (↑).

microgravity (n) the very slight force of gravity (p. 112) that is experienced in a spacecraft. In practice, the contents of a spacecraft are not wholly weightless because small movements of the spacecraft, caused e.g. by the firing of thruster (p. 157) rockets or by movement of the astronaut crew, produce slight accelerations (up to 0.001 g (p. 150) on the Space Shuttle (p. 169)). Therefore microgravity rather than zero gravity is a more exact way to describe conditions in a spacecraft.

satellite

space probe

free fall

e.g. man in lift.
If lift falls freely
man feels weightless

altitude

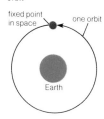

orbit

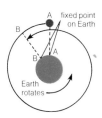

revolution

altitude[2] (n) the height above the surface of the Earth, e.g. of a satellite.

revolution[2] (n) one complete path of a satellite around the Earth, relative to a fixed point on the Earth's surface. In NASA usage, revolutions are counted relative to longitude 80 degrees west, which just touches the Atlantic coast of Florida near Cape Canaveral. Because the Earth is rotating all the time from west to east, a satellite in a west-to-east orbit takes slightly longer to complete one revolution than it does to complete one orbit (an orbit being measured relative to a fixed point in space). Therefore although a spacecraft such as the Space Shuttle (p. 169) may complete 16 orbits in a day, it makes only 15 revolutions of the Earth in the same time.

mission (n) the purpose of a space flight or the spacecraft sent to carry out a duty, e.g. a mission to Mars; the Viking (p. 186) mission; a Space Shuttle (p. 169) mission.

mission control the place from where the activities of a space flight are directed. Mission control for NASA's manned space flights is at the Johnson Space Center in Houston, Texas; for NASA space probes, mission control is at the Jet Propulsion Laboratory in California.

tracking (n) the act of following the course of a rocket or spacecraft by radar or optical means, and of keeping in touch with it by radio. Controllers on Earth communicate with a spacecraft through tracking stations around the world. **track** (v).

Space Tracking and Data Network STDN. The
NASA system for tracking and communicating
with satellites. It consists of 13 ground stations
in North and South America, Europe, and
Australia; in addition, there are several smaller
movable ground stations, and instrumented
aircraft. The ground stations have antennas
(p. 135) from 9 m to 26 m in diameter. The
control centre of the STDN system is NASA's
Goddard Space Flight Center.

Deep Space Network DSN. The NASA system
for tracking and communicating with space
probes. There are three stations of the DSN: at
Goldstone, California; near Madrid; and near
Canberra, Australia. Each of the stations has
an antenna (p. 135) 64 m in diameter, and two
of 26-m diameter. The control centre of the
DSN is NASA's Jet Propulsion Laboratory.

deep space the region of space far from the
Earth, as explored by space probes rather than
by Earth satellites.

ground track

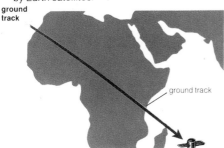

ground track

ground track the path along which a spacecraft
passes directly overhead as seen from the
surface of the Earth.

telemetry (*n*) (1) the flow of instrument readings
from a spacecraft to the ground by means of
radio waves (p. 120); (2) the act of taking and
sending such readings.

real time as something happens, immediate, e.g.
sending back readings from a satellite in real
time rather than recording them and playing
them back later.

acquisition (*n*) the act of receiving something, e.g. the radio signal from a spacecraft, or of finding the position of a rocket or satellite by radar so that it can be tracked. **acquire** (*v*).

uplink (*n*) radio contact from the ground to a spacecraft.

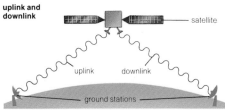

uplink and downlink

downlink (*n*) radio contact from a spacecraft to the ground.

avionics (*n.pl.*) the electronics used in aviation and spaceflight, e.g. for navigation, guidance and communications.

inertial guidance a system that measures and controls the path of a spacecraft. An inertial guidance system uses accelerometers (↓) and gyroscopes (↓) aboard the spacecraft to measure the craft's speed and direction of travel. From these readings a computer works out where the spacecraft is and where it is going. If the craft is not keeping to the intended course, the computer can send out commands to correct it.

inertial measurement unit the system aboard a spacecraft containing accelerometers (↓) and gyroscopes (↓) that provides measurements for navigating the craft.

accelerometer (*n*) an instrument that measures changes in velocity, i.e. accelerations.

gyroscope (*n*) a device used to keep spacecraft on course. A gyroscope consists of a fast-spinning wheel, allowed to swing freely so that its axis always points in the same direction in space, no matter how the spacecraft turns around it. Usually a set of three gyroscopes is used aboard spacecraft to measure changes in direction of travel and attitude (p. 156). **gyro** (*abbr*).

state vector readings concerning the position and movement of a spacecraft, used for navigation. The state vector can either be calculated by instruments aboard the spacecraft or it can be calculated on the ground and sent by radio to the spacecraft's computer.

vacuum (*n*) airless space, e.g. as found above the Earth's atmosphere. In practice, the height at which most satellites orbit is not a complete vacuum; some small amount of air still remains to cause drag (↓).

drag (*n*) the effect caused by the Earth's outer atmosphere on the movement of a satellite. Drag slows the satellite down so that its orbit gets lower, until the satellite finally enters the densest part of the atmosphere and burns up.

attitude (*n*) the position or direction that a spacecraft is pointed in space, e.g. an Earth-pointing attitude.

manoeuvre (*n*) any intentional movement of a spacecraft, e.g. to change its orbit or its attitude (↑). **manoeuvre** (*v*).

roll (*n*) the turning or slow spinning of a rocket or spacecraft around its longest axis, i.e. around the X axis (↓).

pitch (*n*) the up and down movement of the nose of a spacecraft or rocket, i.e. rotation around the Y axis (↓).

yaw (*n*) the side to side (left and right) movement of the nose of a rocket or spacecraft, i.e. rotation around the Z axis (↓).

X axis the longest axis of a spacecraft, i.e. the axis that runs from front to back.

Y axis the horizontal axis in a spacecraft, i.e. the axis that runs from side to side.

Z axis the vertical axis in a spacecraft, i.e. the axis that runs from top to bottom.

attitude

roll, pitch and yaw

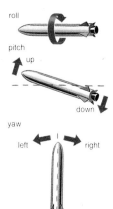

X, Y and Z axes

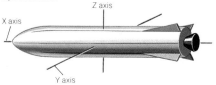

SPACE FLIGHT/MANOEUVRING · 157

translation

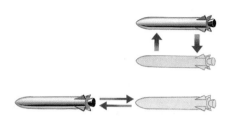

rendezvous

docking

translation (*n*) the movement of a spacecraft in a straight line, whether backwards and forwards, left and right or up and down.

rendezvous (*n*) a close approach of two spacecraft in orbit, especially in preparation for a docking (↓). **rendezvous** (*v*).

docking (*n*) the joining together of two spacecraft in orbit. **dock** (*v*).

thruster (*n*) a small rocket used for controlling the attitude (↑) or stabilization (↓) of a spacecraft, or for making small changes to its position, e.g. during docking (↑).

reaction control system a group of thrusters (↑) on a spacecraft that are used for controlling the spacecraft's attitude (↑) and for making small changes in speed, e.g. during docking (↑).

delta-V a change in velocity, e.g. when changing a spacecraft's orbit.

posigrade (*adj*) (1) of a forward motion in the direction that a spacecraft is travelling; (2) of a small rocket that produces such a movement. A posigrade rocket can be used to raise the height of a satellite's orbit, or to separate two halves of a spacecraft. Posigrade is the opposite of retrograde (p. 14).

stabilization (*n*) the act of keeping the position or the attitude (↑) of a spacecraft steady. **stabilize** (*v*).

spin stabilization the stabilization (↑) of a spacecraft by spinning it, usually around its longest axis. **spin-stabilize** (*v*).

de-spun (*adj*) of a part of a spin-stabilized
(p. 157) spacecraft that is intentionally kept
from spinning, e.g. the radio dish used for
communicating with the Earth.

three-axis stabilization the stabilization (p. 157)
of a spacecraft along each of its three axes – X
axis (p. 156), Y axis (p. 156) and Z axis (p. 156).
A spacecraft that is three-axis stabilized is not
spinning. Small rocket thrusters (p. 157) are
used to keep the spacecraft steady and pointing
in the right direction.

solar panel a frame covered with solar cells (↓),
fixed to a satellite or space probe to provide
electrical power.

solar panel

solar cell a device that turns the energy of
sunlight into electricity. Solar cells are usually
made of a thin flat piece of silicon.

fuel cell a device for producing electricity, usually
used aboard manned spacecraft. In a fuel cell,
atoms (p. 116) of hydrogen and oxygen are
made to join together to produce molecules
(p. 116) of water. In the process, an electric
current is produced. The water made in a fuel
cell is pure and is drunk by the astronauts.

interplanetary (*adj*) (1) of the space between the
planets; (2) of a spacecraft that travels from the
Earth to another planet.

mid-course correction a small change made to
the trajectory (p. 16) of a space probe by
changing its speed, to improve its aim. Several
mid-course corrections may be needed to the
flight path of a space probe to make certain
that it reaches the precise point intended.

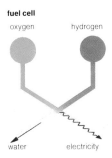

fuel cell

SPACE FLIGHT/RE-ENTRY · 159

re-entry (*n*) the act of coming back into the Earth's atmosphere from space. Friction with the atmosphere during re-entry produces heat that will burn the spacecraft up if it is not protected by a heat shield (p. 160). **re-enter** (*v*).

re-entry corridor the narrow path along which a spacecraft can safely re-enter (↑) the Earth's atmosphere. If the angle of entry is too steep, the heat produced by friction will be so great that even the heat shield (p. 160) will melt away and the craft will burn up. If the angle of entry is too small, the spacecraft will pass through the atmosphere and out into space again.

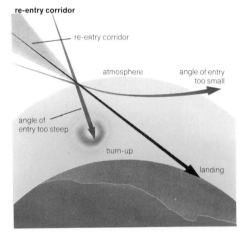

re-entry corridor

decay

decay (*n*) the slow change in size of a satellite's orbit caused by the drag (p. 156) of the Earth's atmosphere. Air drag slows a satellite down, so that its orbit gets lower. The satellite continues to move closer to the Earth until it reaches the dense lower part of the atmosphere and burns up. A satellite that has burned up like this is said to have decayed.

deorbit (*v*) to leave orbit, or to slow a spacecraft down so that it re-enters (↑) the atmosphere, e.g. to make a deorbit burn with retro-rockets (p. 160).

retro-rocket (*n*) a rocket engine that fires to slow a spacecraft down. Retro-rockets can be used to bring a satellite back from orbit around the Earth; to make a soft landing; or to slow a space probe so that it is captured (p. 19) by the Moon or a planet.

retro-fire the firing of retro-rockets (↑) for the return of a spacecraft to Earth.

heat shield a coating on the outside of a spacecraft that protects the craft from the heat produced by friction during re-entry (p. 159). Some heat shields are made of a plastic material that slowly burns away during re-entry (*see* ablation p. 66). The heat shielding on the Space Shuttle (p. 169) consists of tiles and blankets made of low-density silica which do not ablate.

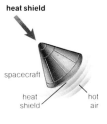

decelerate (*v*) to slow down, or be slowed down.
deceleration (*n*) negative acceleration.
entry interface the point at which a spacecraft enters the dense lower part of the Earth's atmosphere and would begin to burn up if it were not protected by a heat shield (↑). In NASA usage, entry interface is at a height of 400,000 ft, i.e. about 120 km or 75 miles.

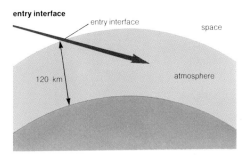

radio blackout the loss of radio contact with a spacecraft during its re-entry (p. 159), caused by the hot ionized (p. 116) gases around the spacecraft which prevent radio waves (p. 120) from getting through. During the re-entry of the Space Shuttle (p. 169), the radio blackout lasts for about 15 minutes.

SPACE FLIGHT/LANDING · 161

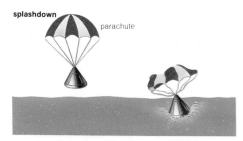

splashdown (*n*) the landing of a spacecraft in the sea. The American manned spacecraft Mercury (p. 166), Gemini (p. 166) and Apollo (p. 164) landed in this way.

parachute (*n*) a large piece of cloth or other material, usually rounded in shape, that allows an object to fall slowly through the atmosphere. All manned spacecraft other than the Space Shuttle (p. 169) have landed under parachutes.

recovery (*n*) the act of bringing a spacecraft back to Earth without damage, or of finding it once it has returned to Earth. **recoverable** (*adj*).

roll reversal an S-shaped turn to left and right made by the Space Shuttle (p. 169) to help slow itself down as it descends through the atmosphere.

crossrange (*n*) the amount that the Space Shuttle (p. 169) can manoeuvre to the left and right of its entry path, i.e. up to 2,000 km.

tactical air navigation a navigation system used by the Space Shuttle (p. 169) after leaving the radio blackout (↑). The system consists of a number of ground stations that send out radio beams, which the Shuttle receives and uses to work out its position and direction of travel.

terminal area energy management the system of last corrections to the speed and height of the Space Shuttle (p. 169), starting about 6 minutes before touchdown (p. 162), to make sure that it will reach the runway.

heading alignment cylinder an imaginary circle 20,000 ft (6 km) in radius, around which the Space Shuttle (p. 169) makes its last turn to line itself up on the runway for landing.

glideslope and touchdown

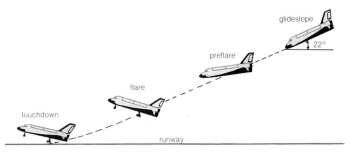

glideslope (*n*) the sloping path along which the Space Shuttle (p. 169) descends towards the runway. The Shuttle's glideslope has an angle of 22 degrees to the ground.

preflare (*n*) the act of levelling out the glideslope (↑) of the Space Shuttle (p. 169) from 22 degrees to 1.5 degrees. This is done about 30 seconds before touchdown (↓).

flare[2] (*n*) the act of raising the nose of the Space Shuttle (p. 169) in readiness for landing. This is done about 15 seconds before touchdown (↓).

touchdown (*n*) the moment that a spacecraft lands on the surface of a planet or moon, e.g. when the wheels of the Space Shuttle (p. 169) touch the runway.

runway (*n*) the long, flat surface from which aeroplanes take off and on which they land. The Space Shuttle (p. 169) has a special landing runway 4.5 km long and 91 metres wide.

astronaut (*n*) any person who travels in space.

cosmonaut (*n*) the Soviet name for an astronaut.

crew (*n*) a group of astronauts who fly together aboard a spacecraft.

simulator (*n*) a device on the ground in which astronauts train for their spaceflight. A simulator is built to be as much like the real spacecraft as possible. In a simulator astronauts practise the work they must do during their flight, and they also learn what to do if anything goes wrong. **simulate** (*v*), **simulation** (*n*).

capsule (*n*) an old name for a manned spacecraft.

astronaut in spacesuit

SPACE FLIGHT/MANNED FLIGHT · 163

hatch

Apollo spacecraft

neutral body position

airlock

1 spacecraft filled with air astronaut in spacesuit

2 astronaut opens hatch and enters airlock

3 astronaut lets air out of airlock

4 astronaut opens outer hatch and moves into space

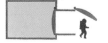

hatch (n) a door for getting in and out of a spacecraft. The hatch fits tightly to keep the air inside the spacecraft.

life support the system that keeps astronauts alive in their spacecraft or in a spacesuit (p. 164). The life-support system provides air conditioning, i.e. it supplies air for the astronauts to breathe, it removes waste products such as carbon dioxide, and it controls the air pressure, temperature and humidity (dampness).

consumable (n) any of the supplies inside a manned spacecraft that are used up during the flight, e.g. air, water, food and fuel.

partial pressure the pressure of one of the gases in a spacecraft's atmosphere, e.g. the air inside the Space Shuttle (p. 169) consists of nitrogen and oxygen; together their pressure adds up to 100 kilopascals (sea level pressure), but the pressure of the oxygen alone is 22 kilopascals, i.e. the oxygen partial pressure is 22 kilopascals.

neutral body position the position that the human body takes up in weightlessness (p. 152), i.e. with arms and legs raised and bent.

space sickness a feeling of sickness felt by astronauts during their first few days in space. It is similar to travel sickness on Earth, and it is caused by the body's reaction to weightlessness (p. 152).

airlock a part of a spacecraft through which astronauts can pass out into space and return again, without having to let all the air out of the main spacecraft. To use an airlock, the astronaut first gets into it and closes the hatch behind him. After checking that his spacesuit (p. 164) is working properly, he lets the air out of the airlock and opens the outer hatch. The astronaut can then move out into space. To return, he gets into the airlock, closes the outer hatch, lets air into the airlock, and then opens the inner hatch into the main spacecraft.

extravehicular activity EVA. The act of moving and working in a spacesuit (p. 164) outside a spacecraft. Extravehicular activity can take place in orbit around the Earth, or it can mean walking on the Moon. Also known as **spacewalk**.

neutral buoyancy tank a large tank of water in which astronauts can practise extravehicular activity (p. 163). The astronauts float in the water so that they feel almost weightless, i.e. the water tank simulates (p. 162) the conditions of weightlessness (p. 152) to be found in space.

spacesuit (n) a special garment that an astronaut wears to protect him from the airless conditions of space. A spacesuit is like a personal spacecraft. The spacesuit provides air for the astronaut to breathe, it protects him from freezing or boiling, and it contains a radio for him to talk. Oxygen for the spacesuit can be provided either through a line from the main spacecraft, or from a pack that the astronaut wears on his back. On the Space Shuttle (p. 169), astronauts wear spacesuits only if they have to go outside the spacecraft. In most earlier spacecraft, astronauts wore spacesuits during launch (p. 148), docking (p. 157) and landing, but took the suits off at other times.

portable life-support system PLSS. A pack on the back of a spacesuit (↑) that provides life support (p. 163) and communications for astronauts while walking on the Moon or working outside the Space Shuttle (p. 169).

extravehicular mobility unit EMU. The complete spacesuit (↑) and portable life-support system (↑) worn by astronauts while on the Moon or while working outside the Space Shuttle (p. 169).

manned manoeuvring unit MMU. A jet-propelled (p. 141) pack used by astronauts for moving around outside the Space Shuttle (p. 169). It fits onto the back of the astronaut's spacesuit and is pushed along by jets of nitrogen gas. The astronaut flies the unit by means of hand controls.

Apollo (n) an American manned spacecraft that carried a crew of three. Apollo spacecraft took men to the Moon and also to the Skylab (p. 168) space station (p. 167). In 1975 an Apollo spacecraft docked (p. 157) in orbit with a Soviet Soyuz (p. 167) spacecraft.

module (n) any self-contained part that goes together with others to make up a complete spacecraft or rocket. **modular** (adj).

portable life-support system

Apollo

command module
service module
astronauts
fuel tanks
service module main engine

command module CM. The part of a manned spacecraft, especially Apollo (↑), that contains the crew and the main controls. The Apollo command module was conical in shape, 3.9 m across at its widest and 3.5 m high. In flight, it was joined to the service module (↓).

service module SM. The part of a manned spacecraft, especially Apollo (↑), that contains supplies of air, water and electricity for the command module (↑), and engines for manoeuvring. The Apollo service module was a cylinder, 3.9 m in diameter and 7.5 m long. It is jettisoned (p. 151) before re-entry (p. 159).

service propulsion system SPS. The large rocket engine at the back of the Apollo (↑) service module. It was used for major course changes, and for putting the spacecraft into orbit around the Moon and for sending it back to Earth again.

lunar module LM. That part of the Apollo (↑) spacecraft in which two astronauts landed on the Moon. The Apollo lunar module came in two halves: a descent stage which contained four legs and the retro-rockets (p. 160) for landing; and the ascent stage in which the astronauts rode and in which they took off from the Moon's surface to rejoin the command module (↑). The total height of the lunar module when on the Moon was 7 m.

lunar module

166 · SPACE FLIGHT/MANNED FLIGHT

lunar roving vehicle

lunar roving vehicle LRV. An electrically powered car in which Apollo (p. 164) astronauts rode while on the surface of the Moon. Lunar rovers were used on the flights of Apollo 15, 16 and 17.

Gemini (*n*) an American manned spacecraft with a crew of two. Astronauts aboard Gemini craft practised rendezvous (p. 157) and docking (p. 157) in preparation for the Apollo (p. 164) flights to the Moon. A series of 10 manned Gemini flights were made in 1965 and 1966.

Mercury[2] (*n*) an American manned spacecraft that carried one astronaut. The first six American manned spaceflights were made in Mercury capsules (p. 162); the first two were only sub-orbital (p. 151).

Gemini

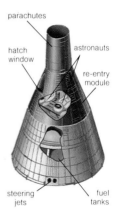

Mercury

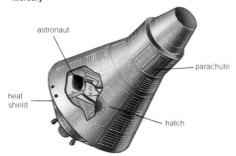

SPACE FLIGHT/MANNED FLIGHT · 167

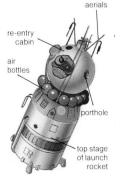

Vostok
aerials
re-entry cabin
air bottles
porthole
top stage of launch rocket

Vostok (*n*) a Soviet manned spacecraft that carried one cosmonaut. Vostok was a sphere 2.3 m in diameter. The first human to fly in space, Yuri Gagarin, made his flight in Vostok 1 on April 12, 1961.

Voskhod (*n*) an improved Vostok (↑) spacecraft that could carry more than one cosmonaut. Voskhod 1 in October 1964 carried a crew of three. Voskhod 2 in March 1965 carried two men, one of whom, Alexei Leonov, left the spacecraft through an airlock (p. 163) to perform the first extravehicular activity (p. 163).

Soyuz (*n*) a Soviet manned spacecraft that can carry up to three cosmonauts. On its first flight in April 1967 it crashed after re-entry (p. 159), killing the single cosmonaut on board, Vladimir Komarov. Soyuz spacecraft are now used to take crews up to the Salyut (p. 168) space station (↓), although in the past they have made flights on their own. Since 1980 an improved form of Soyuz, called *Soyuz T*, has been used.

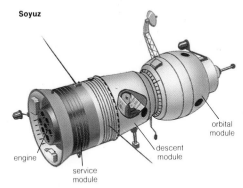

Soyuz
orbital module
descent module
engine
service module

space station a large spacecraft in which crews of astronauts and cosmonauts can live and work for weeks or months at a time. In a space station astronauts can observe the Earth and sky and do scientific work that makes use of the condition of weightlessness (p. 152). *See* Skylab (p. 168) and Salyut (p. 168).

168 · SPACE FLIGHT/SPACE STATIONS

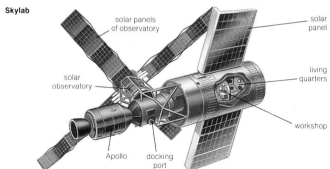

Skylab

Skylab (*n*) an American space station (p. 167), made from the top stage (p. 141) of a Saturn V (p. 174) rocket. Skylab was 26m long and weighed 75 tonnes, the largest and heaviest payload (p. 142) ever put into orbit. It was launched in May 1973 and re-entered (p. 159) in 1979. Three crews, each of three astronauts, manned the station for up to three months.

Salyut (*n*) a series of Soviet space stations (p. 167) made from the top stage (p. 141) of Proton (p.177) rockets. Salyut is 15 m long and weighs 19 tonnes; additional modules (p. 164) can be docked (p.157) to Salyut to make it larger. Salyut 1 was launched in April 1971. Several others have followed. Crews of cosmonauts have made flights of six months or more inside Salyut.

Mir (*n*) an improved form of Soviet space station (p. 167), first launched in 1986.

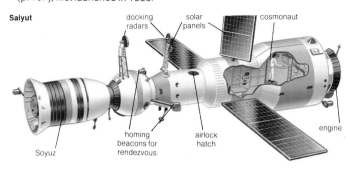

Salyut

SPACE FLIGHT/SPACE SHUTTLE · 169

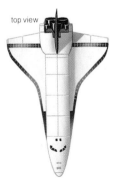

top view

front view

rear view

bottom view

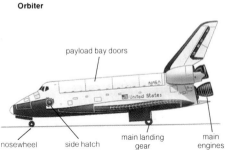

Orbiter

payload bay doors

nosewheel   side hatch   main landing gear   main engines

Space Shuttle an American launch vehicle (p. 141) that takes off like a rocket but which lands on a runway like an aeroplane. Unlike previous rockets, it can be used again. It goes into orbit about 300 km above the Earth; it cannot go any higher than 1,000 km. The Shuttle can carry into space a payload (p. 142) of up to 29 tonnes, and bring back to Earth up to 14 tonnes. The total weight of the Shuttle at launch (p. 148) is about 2,000 tonnes. Its engines produce a thrust (p. 145) of nearly 3,000 tonnes at launch.

Space Transportation System STS. The NASA launch (p. 148) system of which the Space Shuttle (↑) is the main part.

Orbiter (*n*) the part of the Space Shuttle (↑) that goes into orbit around the Earth and which comes back to land on a runway. It carries a crew of up to seven people, who ride in the Orbiter's front part. The Orbiter is 37.2 m long, 23.8 m wide across its wings, and weighs 75 tonnes empty. It has three engines that fire to put the craft into orbit; their fuel is supplied from the External Tank (p. 170). But no engines are used when landing the Orbiter. The Orbiter is covered with a thermal protection system (↓).

thermal protection system the heat shield (p. 160) of the Space Shuttle (↑) Orbiter (↑). The thermal protection system is made mostly of pieces of low-density silica stuck to the Orbiter, but carbon is used in the areas that get hottest, i.e. the nose and the edges of the wings.

170 · SPACE FLIGHT/SPACE SHUTTLE

External Tank ET. A large tank joined to the
Space Shuttle (p. 169) Orbiter (p. 169) at launch
(p. 148). It contains 2 million litres of liquid
hydrogen and liquid oxygen propellants (p. 143)
for the Space Shuttle's main engines. All the
propellants are used up after 8.5 minutes, and
the empty External Tank drops away to burn up
in the atmosphere. It is the only part of the
Space Shuttle system not intended to be
reused.

Solid Rocket Booster SRB. Either of two booster
(p. 142) rockets using solid propellants (p. 143)
that are joined to the side of the Space
Shuttle's (p. 169) External Tank (↑) at launch
(p. 148). Each produces a thrust (p. 145) of
1,200 tonnes to help lift the Shuttle off the
ground. The Solid Rocket Boosters burn out
after 2 minutes, at a height of about 45 km.
They fall away under parachutes (p. 161) into
the ocean and are recovered to be used again.

orbital manoeuvring system OMS. Two rocket
engines at the back of the Space Shuttle
(p. 169) Orbiter (p. 169) that fire to put the
Shuttle into orbit after the main engines have
burned out and the External Tank (↑) has
dropped away. The orbital manoeuvring system
uses its own hypergolic (p. 145) propellants
(p. 143). The orbital manoeuvring system
engines are used to change the Shuttle's orbit
in space, and they also act as
retrorockets (p. 160) to bring the Shuttle back
to Earth.

External tank

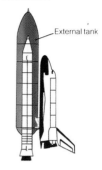

Solid Rocket Booster

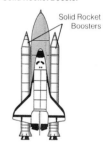

orbital manoeuvring system

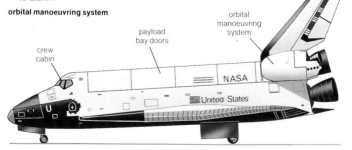

SPACE FLIGHT/SPACE SHUTTLE · 171

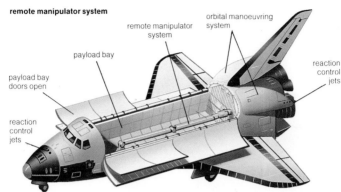

flight deck the part of the Space Shuttle (p. 169) Orbiter (p. 169) that contains the controls for flying the craft. At launch (p. 148) and landing, up to four astronauts are seated on the flight deck, including the pilot and co-pilot. The flight deck also contains the controls for working the remote manipulator system (↓) and for launching satellites from the payload bay (↓).

mid deck the part of the Space Shuttle (p. 169) Orbiter (p. 169) in which the crew live, eat and sleep. The mid deck is below the flight deck (↑). At launch (p. 148) and landing, up to three astronauts can be seated on the mid deck.

payload bay the part of the Space Shuttle (p. 169) Orbiter (p. 169) that carries satellites, scientific equipment or the Spacelab (p. 172) space station (p. 167). It is 18.3 m long and 4.6 m wide. It has two doors that swing open once the Shuttle is in orbit; the doors are closed again before re-entry (p. 159). Also known as **cargo bay**.

remote manipulator system RMS. A mechanical arm used for putting satellites into orbit from the payload bay (↑) of the Space Shuttle (p. 169) and for catching hold of other satellites to bring them into the payload bay, e.g. for repair. The remote manipulator arm is controlled by astronauts on the flight deck (↑) of the Orbiter (p. 169). When not in use, the arm is fixed to the inside of the payload bay.

172 · **SPACE FLIGHT**/SPACELAB

Spacelab

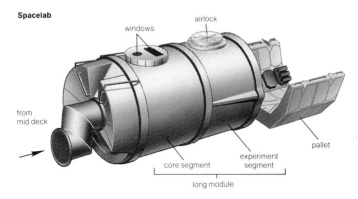

Spacelab (*n*) a small space station (p. 167) that flies in the payload bay (p. 171) of the Space Shuttle (p. 169). It was built for NASA by the European Space Agency. Spacelab consists of two parts: the pressurized module (p. 164) in which astronauts can work, and pallets (↓) in open space to which instruments and equipment can be fixed. The pressurized module and the pallets can be flown together, or on their own.

short module the core segment (↓) of Spacelab (↑) on its own.

long module the core segment (↓) and the experiment segment (↓) of Spacelab (↑) joined together.

core segment part of the Spacelab (↑) pressurized module (p. 164) that contains the control equipment for Spacelab. It is 4 m wide and 2.7 m long. It can be flown on its own in the payload bay (p. 171) of the Space Shuttle (p. 169); it is then known as the short module (↑). The experiment segment (↓) can be joined to it to form the long module (↑).

experiment segment part of the pressurized module (p. 164) of the Spacelab (↑) space station (p. 167). It contains equipment for carrying out experiments, and gives astronauts extra room in which to work. It is 4 m wide and 2.7 m long. It is joined to the core segment (↑) to form what is known as the long module (↑).

SPACE FLIGHT/SPACE SHUTTLE · 173

pallet (*n*) a part of the Spacelab (↑) space station (p. 167) that is open to the vacuum of space. Each is a U-shaped frame to which equipment and instruments can be fixed. Astronauts in Spacelab can control the equipment on the pallets. As many as five can be used at a time. When the pallets are flown on their own, i.e. without the pressurized module (p. 164), the equipment on them is controlled by astronauts from the Orbiter's (p. 169) flight deck (p. 171).

igloo (*n*) a pressurized container that handles services such as electrical supplies and communications for the equipment on the Spacelab (↑) pallets (↑) when the pallets are flown on their own.

mission specialist a NASA astronaut who flies on the Space Shuttle (p. 169) and who is in charge of the scientific work to be carried out during the flight, including putting satellites into orbit from the payload bay (p. 171). They use the remote manipulator system (p. 171), and they perform any extravehicular activity (p. 163).

payload specialist someone, e.g. a scientist or engineer, who is not a full-time astronaut but who flies on the Space Shuttle (p. 169) to perform a particular experiment or to use a particular piece of equipment.

getaway special GAS. A small payload (p. 142) flown in a container in the payload bay (p. 171) of the Space Shuttle (p. 169). Getaway specials must be self contained, i.e. they must have their own electrical supplies and they must work on their own without the help of astronauts. They must weigh under 91 kg and take up less than 0.14 cubic metres. Getaway specials are flown whenever there is room in the payload bay.

Enterprise (*n*) a Space Shuttle (p. 169) Orbiter (p. 169) used by astronauts to practise landing after it had been taken up into the air on the back of an aeroplane. It did not have real engines or a thermal protection system (p. 169).

Columbia (*n*) the first Space Shuttle (p. 169) Orbiter (p. 169) to fly in space. It made its first flight on April 12, 1981. It also carried the first Spacelab (↑) in November 1983.

174 · SPACE FLIGHT/ROCKETS

Saturn V

Challenger (n) the second Space Shuttle (p. 169) Orbiter (p. 169) to fly in space. It made its first flight in 1983. It was destroyed in an explosion on its tenth launch (p. 148) on January 28, 1986.

Discovery (n) the third Space Shuttle (p. 169) Orbiter (p. 169) to fly in space. It made its first flight in 1984.

Atlantis (n) the fourth Space Shuttle (p. 169) Orbiter (p. 169) to fly in space. It made its first flight in 1985.

Ariane (n) a three-stage rocket built by the European Space Agency for launching (p. 148) satellites, especially into geostationary orbit (p. 23), and the Giotto (p. 187) space probe. The most powerful forms of Ariane use solid-propellant strap-on boosters (p. 142) on the first stage (p. 141).

Atlas (n) an American rocket that launched (p. 148) Mercury astronauts into orbit around the Earth. With upper stages (p. 141) such as the Agena (p. 176) and Centaur (p. 176), the Atlas is used to launch satellites and space probes.

Delta (n) an American rocket used to launch (p. 148) many kinds of satellites. A Delta rocket can have two or three stages (p. 141), with up to nine solid-propellant strap-on boosters (p. 142) joined to its first stage.

Saturn IB a two-stage American rocket used for putting Apollo (p. 164) spacecraft into orbit around the Earth, e.g. to reach the Skylab (p. 168) space station (p. 167). With Apollo on top, the Saturn IB stood 68 m tall; at launch (p. 148) its engine thrust (p. 145) was 740 tonnes. The Saturn IB was last used in July 1975 to send an Apollo into orbit to dock with a Soviet Soyuz (p. 167) spacecraft.

Saturn V a three-stage American rocket, the largest and most powerful ever built. Saturn V rockets were used to send Apollo (p. 164) spacecraft to the Moon. With Apollo on top, the Saturn V stood 111 m tall; its engine thrust (p. 145) at launch (p. 148) was 3,500 tonnes. It was last used in May 1973 to launch the Skylab (p. 168) space station (p. 167), which was made out of its top (third) stage (p. 141).

SPACE FLIGHT/ROCKETS · 175

Scout (*n*) an American solid-propellant rocket of four or five stages (p. 141), used for launching (p. 148) small scientific satellites.

Thor (*n*) an American rocket used in the 1960s and 1970s to launch (p. 148) satellites, usually with Agena (p. 176) or Delta upper stages (p. 141). In improved form it became known as the Delta (↑).

Titan (*n*) an American rocket used for launching (p. 148) many kinds of satellites and space probes. Two-stage Titan rockets launched the manned Gemini (p. 166) spacecraft. With upper stages (p. 141) and two large solid-propellant strap-on boosters (p. 142) Titan rockets have launched many military satellites. Titan rockets with strap-on boosters and a Centaur (p. 176) upper stage were used to launch the Viking (p. 186) and Voyager (p. 187) space probes.

rockets

176 · SPACE FLIGHT/ROCKETS

Agena (*n*) an American upper stage (p. 141) that was used with rockets such as the Atlas (p.174) and Titan (p. 175) for launching (p. 148) many kinds of satellites and some space probes.

Centaur (*n*) an American upper stage (p. 141), used for launching (p. 148) space probes and heavy satellites with Atlas (p. 174) and Titan (p. 175) rockets.

Payload Assist Module PAM. A solid-propellant rocket stage (p. 141) that is fitted to satellites launched (p. 148) from the Space Shuttle's (p. 169) payload bay (p. 171). It sends the satellites into geostationary orbit (p. 23). There are two different Payload Assist Modules: the more powerful *PAM-A* is used for satellites formerly launched by Atlas-Centaur rockets; the *PAM-D* is used for satellites formerly launched by Delta (p. 174) rockets. The Payload Assist Module is spin-stabilized (p. 157); also known as **spinning solid upper stage** (SSUS).

Inertial Upper Stage IUS. A two-stage solid-propellant rocket that sends large satellites carried by the Space Shuttle (p. 169) into geostationary orbit (p. 23). The Inertial Upper Stage is three-axis stabilized (p. 158).

A-type the name given in the West to the Soviet rocket that launched (p. 148) Sputnik 1 (p. 178) and that, with added upper stages (p. 141), has launched the Vostok (p. 167), Voskhod (p. 167) and Soyuz (p. 167) spacecraft, as well as many satellites and space probes. It has four liquid-propellant strap-on boosters (p. 142) around a central sustainer (p. 142) engine.

B-type the name given in the West to a Soviet rocket of two stages (p. 141), used for launching (p. 148) small scientific satellites. The smallest Soviet launch rocket.

C-type the name given in the West to a two-stage Soviet launch (p. 148) rocket, used for orbiting scientific and military satellites.

D-type the name given in the West to a Soviet launch (p. 148) rocket of two or more stages (p. 141). The D-type rocket is used to launch space probes and the Salyut (p. 168) space station (p. 167). Also known as **Proton rocket**.

SPACE FLIGHT/ROCKETS · 177

F-type the name given in the West to a three-stage Soviet rocket used for military launches (p. 148), including possible orbital weapons systems. (NB: there is no Soviet E-type rocket).
G-type the name given in the West to a possible future Soviet launch (p. 148) rocket, similar in size and power to the American Saturn V (p. 174). The G-type rocket is expected to launch large space stations (p. 167). It may also send cosmonauts to the Moon.
Proton rocket = D-type (↑).
Vostok rocket = A-type (↑).

rockets Soviet

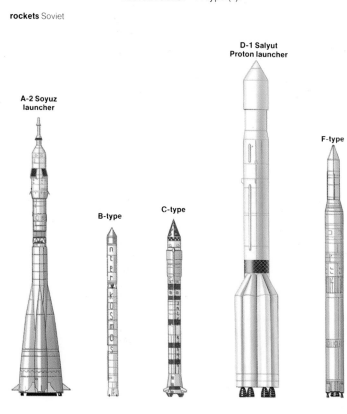

178 · SATELLITES

Sputnik

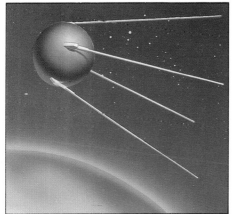

Sputnik (*n*) a series of ten Soviet Earth satellites. Sputnik 1 was the first artificial satellite (p. 152); it was launched (p. 148) on October 4, 1957. Sputnik 2 a month later carried the dog Laika into orbit. Several later Sputniks were actually test flights of unmanned Vostok (p. 167) spacecraft. The Sputniks were succeeded by the Cosmos (↓) series.

Explorer (*n*) a continuing series of American scientific satellites. Explorer 1 was the first U.S. Earth satellite, launched (p. 148) on January 31, 1958. Since then, Explorer satellites have made many scientific discoveries.

Vanguard (*n*) the second American Earth satellite. Vanguard 1 was launched (p. 148) into orbit on March 17, 1958, by a rocket also called Vanguard. Two other Vanguard satellites followed, but the series finished in 1959.

Cosmos (*n*) a continuing series of Soviet Earth satellites. Cosmos 1 was launched (p. 148) on March 16, 1962. Many others have followed, at rates of up to 100 a year. Cosmos satellites are launched for scientific and military purposes. The name Cosmos is sometimes given to other spacecraft that have failed to work.

Explorer 1

communications satellite

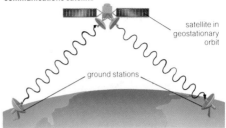

communications satellite a satellite through which messages pass by radio waves (p. 120) from one place on Earth to another. Communications satellites carry telephone and telex messages, TV and radio programmes, and computer data. Most communications satellites are put into geostationary orbit (p. 23). The entire world is served by a series of communications satellites called Intelsat (p. 180), but an increasing number of countries are putting up satellites for their own communications. **comsat** (*abbr*).

Telstar (*n*) a communications satellite (↑) that carried the first live television pictures across the Atlantic Ocean, in July 1962. Unlike present-day communications satellites, which are in geostationary orbit (p. 23), Telstar was in a low orbit; it moved across the sky and so it was in view of any two ground stations for only a limited period of time.

Telstar

Intelsat series

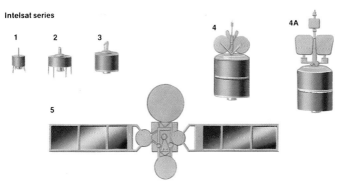

Intelsat (*n*) a series of communications satellites (p. 179) that cover the entire world. Intelsat satellites are placed in geostationary orbit (p. 23) above the Atlantic, Pacific and Indian Oceans. The first Intelsat, called Early Bird, was put into orbit above the Atlantic Ocean in 1965; it could carry 240 telephone calls or one TV programme. The latest Intelsat satellites can carry 15,000 telephone calls and two TV channels. Even larger satellites are planned. The satellites are owned by an international company, also called Intelsat, which has its offices in Washington, D.C.

direct-broadcast satellite a communications satellite (p. 179) that is so large and powerful it can beam signals direct to small aerials (p. 135) on the roofs of offices, homes and in gardens.

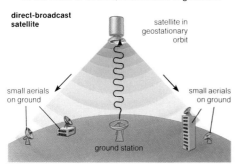

Tracking and Data Relay Satellite TDRS. A NASA satellite in geostationary orbit (p. 23) used for the tracking of, and communication with, satellites in lower orbits, including the Space Shuttle (p. 169). Radio messages flow through the Tracking and Data Relay Satellite to its ground station at White Sands, New Mexico. TDRS satellites will be stationed in orbit over the Atlantic and Pacific Oceans.

navigation satellite a satellite that allows ships at sea, aircraft in the air and people on land to find their exact positions. The satellite gives out radio signals that tell where it is in space. These signals are picked up by a small receiver on a ship, etc, and a small computer works out the receiver's position with respect to the satellite.

navigation satellite

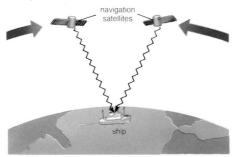

Transit (*n*) a series of American navigation satellites (↑). Transit 1, the world's first navigation satellite, was launched (p. 148) in April 1960. Transit satellites orbited the Earth from pole to pole at heights of about 1,000 km.

Navstar (*n*) a series of 18 American navigation satellites (↑) that make up the Global Positioning System (GPS). Navstar satellites allow navigators to find their position on Earth to about 10 metres, their velocity to within 0.1 metres per second, and time to within a millionth of a second. The Navstars orbit the Earth every 12 hours at a height of 20,000 km.

182 · SATELLITES

weather satellite a satellite that photographs the clouds over the Earth, and measures the temperature and dampness of the atmosphere. It gives warning of dangerous storms, and allows scientists to tell what the weather will be in the near future. Some weather satellites are in geostationary orbit (p. 23) above the equator; others are in lower orbits that take them from pole to pole every 100 minutes.

Tiros (n) a series of American weather satellites (↑). The name stands for *T*elevision and *InfraR*ed *O*bservation *S*atellite. The first Tiros satellite was launched (p. 148) in April 1960. In 1978 a new series of improved satellites was started, called Tiros N, which cover the Earth from low polar orbit (p. 23) four times daily.

Nimbus (n) a series of seven NASA satellites launched (p. 148) between August 1964 and October 1978 that tried out new equipment for weather satellites (↑). The Nimbus satellites were put into polar orbit (p. 23) about 1,000 km high.

Meteosat (n) a weather satellite (↑) built and launched (p. 148) by the ESA into geostationary orbit (p. 23) above the Atlantic Ocean.

weather satellite

photograph of clouds from weather satellite

Meteosat system

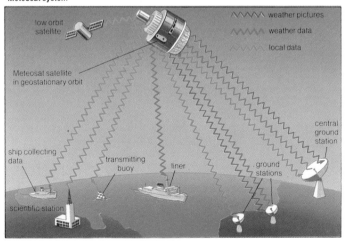

remote sensing

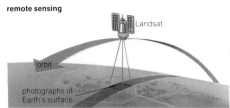

remote sensing the act of studying the Earth's surface for scientific purposes from space by photographs, radar, etc. Remote sensing satellites can be used e.g. to make maps; to find new sources of oil and valuable metals; to study seas, rivers, lakes, deserts, rocks and soil; to watch the growth of crops and forests.

Landsat (n) a series of five NASA remote sensing (↑) satellites. Landsat 1 was launched (p. 148) in July 1972. The Landsats orbit at a height of 1,000 km, circling from pole to pole 14 times a day and continually taking photographs and making other measurements of the Earth beneath them as they do so.

reconnaissance satellite a satellite that photographs the Earth in detail for military purposes. Reconnaissance satellites can see objects as small as people and vehicles on the ground. These satellites can show how many tanks, aeroplanes, missiles, etc, each country has and where they are placed. Also known as **spy satellite**.

ferret satellite a satellite that listens to the radio transmissions of other countries, especially the communications of governments and armed forces.

solar power satellite SPS. A possible satellite of the future that would collect solar energy in space and beam it back to Earth in the form of short-wavelength radio waves (p. 120), where it would be turned into electricity. One solar power satellite could, in theory, produce as much electricity as several normal power stations. Solar power satellites would be placed in geostationary orbit (p. 23). They would be very big, very heavy and costly to build.

Luna (*n*) a series of Soviet Moon probes. Luna 2 was the first constructed object to hit the Moon when it crashed on the surface in September 1959 (Luna 1 had missed). Luna 3 took the first photographs of the far side of the Moon in October 1959. Luna 9 was the first probe to land on the Moon and send back pictures, in February 1966. Three of the Luna series brought back pieces of rock and soil from the surface of the Moon, and two carried unmanned Moon cars called Lunokhod (↓) to the Moon.

Ranger (*n*) a series of American Moon probes. Rangers 7, 8 and 9 sent back close-up photographs of the Moon in 1964 and 1965.

Lunar Orbiter a series of five American probes that went into orbit around the Moon, in 1966 and 1967. They photographed the entire Moon in detail, and took close-up looks at possible landing places for the Apollo (p. 164) astronauts.

Surveyor (*n*) a series of seven American probes meant to land on the Moon. Surveyors 2 and 4 crashed, but the rest were successful. They sent back pictures from the Moon's surface and studied the Moon's rocks.

Surveyor

Lunokhod

Lunokhod (n) either of two Soviet unmanned Moon cars that landed on the Moon in 1970 and 1973. They were driven around by radio command from controllers on Earth. They took pictures of the Moon's surface.

Zond (n) a series of four Soviet space probes. Zond 3 in 1965 took photographs of the far side of the Moon, but the others did not work.

Mariner (n) a series of ten American planetary probes. Mariners 1, 3 and 8 were launch (p. 148) failures. Mariner 2 was the first probe to reach Venus, which it flew past in December 1962. Mariner 5 also flew past Venus. Mariner 4 was the first probe to reach Mars; it flew past in 1965, sending back photographs of its surface. Mariners 6 and 7 also flew past Mars, and Mariner 9 went into orbit around Mars. Mariner 10 was the first probe to visit two planets. In 1974 it flew past Venus and then went on to Mercury, where it photographed the planet's cratered (p. 34) surface for the first time.

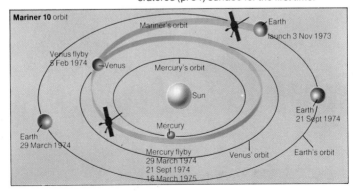

Venera (*n*) a series of Soviet probes to Venus. Venera 7 landed on the surface by parachute (p. 161) in 1970 and sent back the first radio messages from the surface. Later Venera probes took photographs of the surface of Venus.

Mars probes a series of Soviet probes to Mars (for American Mars probes, *see* Mariner (p. 185) and Viking (↓)). Mars 2 and 3 in 1971 and Mars 5 in 1974 sent back some readings from orbit around Mars, but the rest of the series was unsuccessful.

Pioneer (*n*) a series of 11 American space probes. The first four Pioneers, in 1958 and 1959, were meant to reach the Moon, but none were successful. Pioneers 5 to 9 studied conditions in space between the planets. Pioneers 10 and 11 were the first probes to go to Jupiter; Pioneer 11 also went on to look at Saturn. Pioneer 10 became the first constructed object to leave the solar system.

Pioneer-Venus (*n*) two American space probes to Venus in 1978. Pioneer-Venus went into orbit around Venus to study the planet's surface by radar. Pioneer-Venus 2 carried four smaller probes that entered the planet's atmosphere.

Viking (*n*) either of two American space probes to Mars. Each Viking came in two halves: a part that went into orbit around Mars and a part that landed on the surface. Both Vikings got to Mars in 1976. Neither probe found any signs of life.

Viking
Viking probe on Mars

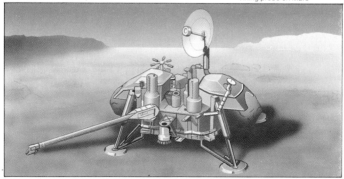

SPACE PROBES · 187

Voyager

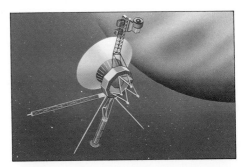

Giotto
Giotto approaching
Halley's comet

Voyager (*n*) either of two American space probes to the outer planets. Voyager 1 flew past Jupiter in 1979 and Saturn in 1980. Voyager 2 followed it a few months later. Voyager 2 was also sent on to reach Uranus in 1986 and Neptune in 1989.

Giotto (*n*) a space probe to Halley's comet (p. 61), built and launched by the European Space Agency. Giotto was launched in July 1985 and reached Halley's comet in March 1986.

Vega² (*n*) either of two Soviet-French probes to Venus and Halley's comet (p. 61). After passing Venus in June 1985, where they dropped instrumented probes into the planet's atmosphere, they flew on to look at Halley's comet, arriving in March 1986.

Galileo (*n*) an American probe to study Jupiter. The Galileo probe comes in two halves: one part to drop into the atmosphere of Jupiter, and the other part to go into orbit to study the planet's clouds and moons.

Galileo
Galileo approaching Jupiter
entry probe dropped off

Orbiting Astronomical Observatory OAO. A series of three NASA satellites that studied ultraviolet (p. 120) light and X-rays (p. 120) coming from stars and galaxies. OAO 1 was launched (p. 148) in April 1966; OAO 2 was launched in December 1968; OAO 3 was launched in August 1972.

Copernicus satellite another name for the third Orbiting Astronomical Observatory (↑), OAO 3. It was named after Nicolaus Copernicus. It carried an 81-cm telescope for spectroscopic (p. 120) observations of the ultraviolet (p. 120) light from stars, and studied X-rays (p. 120) from the possible black hole (p. 79) called Cygnus X-1 (p. 79).

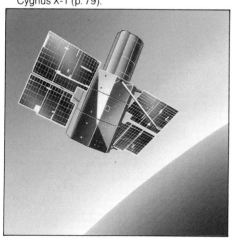

Copernicus satellite

Small Astronomy Satellite SAS. Any of three NASA satellites in the Explorer (p. 178) series. SAS-A, also known as Explorer 42 and Uhuru (↓), was launched (p. 148) in December 1970; it was the first satellite to study X-rays (p. 120) from space. SAS-B (Explorer 48) was launched in November 1972; it was the first satellite to study gamma rays (p. 120) from space. SAS-C (Explorer 53) was launched in May 1975 to examine particular X-ray sources in more detail.

Uhuru (*n*) the name given to the first Small Astronomy Satellite (↑) after it was launched (p. 148). Uhuru is the Swahili word for Freedom.

High Energy Astronomy Observatory HEAO. A series of three NASA satellites that studied X-rays (p. 120) and gamma rays (p. 120) from objects in space. HEAO 1 was launched (p. 148) in August 1977; HEAO 2, also called the Einstein observatory (↓), was launched in November 1978; and HEAO 3 was launched in September 1979.

Einstein observatory another name for the second High Energy Astronomy Observatory (↑), HEAO 2, named after the famous scientist who was born 100 years earlier.

Einstein observatory

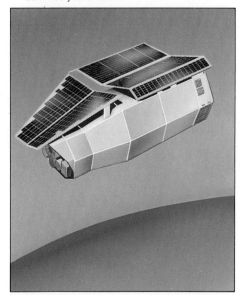

Ariel (*n*) a series of six United Kingdom astronomy satellites, launched (p. 148) by NASA. The first was launched in April 1962, the last in June 1979. The most successful, Ariel V, launched in October 1974, made a complete map of X-ray (p. 120) sources in the sky.

190 · ASTRONOMY SATELLITES

COS-B a satellite of the European Space Agency which studied gamma rays (p. 120) coming from objects in space. COS-B was launched (p. 148) in August 1975.

International Ultraviolet Explorer IUE. A satellite owned jointly by NASA, the United Kingdom and the European Space Agency. IUE contains a telescope with a mirror 45 cm in diameter to study ultraviolet (p. 120) light coming from stars and galaxies. It was launched in January 1978.

International Ultraviolet Explorer

InfraRed Astronomy Satellite IRAS. A satellite owned jointly by NASA, the Netherlands and the United Kingdom. It was put into orbit in January 1983 carrying a telescope with a mirror 60 cm in diameter. It was the first satellite to map the sky at infrared (p. 120) wavelengths (p. 119).

InfraRed Astronomy Satellite

Exosat (*n*) a satellite of the European Space Agency launched (p. 148) in May 1983 to study objects in space that give out X-rays (p. 120).

Orbiting Solar Observatory OSO. A series of eight NASA satellites to study the Sun over one complete cycle of activity. The OSO series were launched between March 1962 and June 1975.

International Sun-Earth Explorer ISEE. A series of three satellites built by NASA and the ESA to study the Sun's effects on the Earth and its magnetosphere (p. 44). ISEE 1 and 2 were launched (p. 148) in October 1977; ISEE 3 followed in August 1978. After its work around the Earth was done, ISEE 3 was sent to examine the tail of comet Giacobini-Zinner in 1985, and to study the effects of the solar wind (p. 32) on Halley's comet (p. 61) in 1985 and 1986.

Solar Maximum Mission SMM. A NASA satellite launched in February 1980 to study the Sun at its greatest activity. After nine months in orbit it went wrong, but was repaired in April 1984 by astronauts from the Space Shuttle (p. 169).

Solar Maximum Mission

Hubble Space Telescope

Hubble Space Telescope the largest telescope ever to be placed in orbit, with a mirror 2.4 m in diameter. It is due for launch by the Space Shuttle (p. 169). It should be able to see objects 10 times smaller and 100 times fainter than telescopes on the ground. The Hubble Space Telescope should help astronomers find out much more about the Universe and the way in which it was formed.

Hipparcos (*n*) a satellite of the European Space Agency planned for launch (p. 148) in 1988. Hipparcos will be used to measure exact star positions for astrometry (p. 89).

GENERAL WORDS IN ASTRONOMY

absolute (*adj*) of some fixed quantity or fact, not measured relative to anything else.

acronym (*n*) a word or name made up from the initial letters of other words, or a shortened form of those words, e.g. NASA. *See* p. 209.

apparent (*adj*) of the way that something appears.

astronomy (*n*) the study of the sky and the objects in it. **astronomer** (*n*), **astronomical** (*adj*).

attraction (*n*) a force that tries to draw objects together. **attract** (*v*).

average (*n*) the middle of a range of quantities. It is calculated by adding up all the quantities and dividing by the number of quantities, e.g. the average of 6, 10 and 20 = 36/3 = 12.

axis (*n*) the line around which something turns, e.g. the axis of the Earth. **axial** (*adj*), **axes** (*pl*).

boundary (*n*) the edge, end or border of something.

celestial (*adj*) of the heavens.

classify (*v*) to arrange things into groups or classes. **classification** (*n*).

cluster (*n*) a group of things close together.

communicate (*v*) to send and receive messages. **communication** (*n*).

cone (*n*) a figure with a round bottom and a pointed top. **conical** (*adj*).

constant (*adj*) of something that is fixed, regular or unchanging.

contingency (*n*) a chance event, especially an accident or failure.

contract (*v*) to get smaller, narrower. **contraction** (*n*).

control (*v*) to guide, govern or aim something, e.g. to control the flight path of a rocket.

converge (*v*) to come together. **convergent** (*adj*).

cosmic (*adj*) of the Universe. **cosmos** (*n*).

crust (*n*) a hard outer surface or skin.

cycle (*n*) a set of events that happen over and over again; the time taken for such events to repeat themselves.

data (*n.pl.*) facts and figures; readings from an instrument.

dense (*adj*) of something that is crowded or pressed tightly together.

density (*n*) the amount of matter in a given volume. The density of an object is calculated by dividing its mass by its volume.

axis

cone

diameter

diverge

filter

red filter lets through red light

blue filter lets through blue light

detail (*n*) small parts or markings of something.
diagram (*n*) a drawing using mostly lines that shows an idea in simple form.
diameter (*n*) the width of an object, especially of a circle or sphere.
dim (*adj*) faint; not very bright. (*v*) to make fainter.
diverge (*v*) to move apart. **divergent** (*adj*).
energy (*n*) the ability of something to do work; power, force.
equation (*n*) a mathematical expression to show that one thing is equal to another; a correction.
exobiology (*n*) the study of possible forms of life off Earth.
expand (*v*) to grow bigger, wider. **expansion** (*n*).
experiment (*n*) something that is done, e.g. scientific work, to see what will happen; an attempt to prove a theory or to discover some new fact. **experiment** (*v*).
explore (*v*) to travel into unknown places to learn about them; to examine at first hand. **exploration** (*n*).
extend (*v*) to reach out, stretch or lengthen.
extraterrestrial (*adj*) of something beyond or off the Earth.
filter (*n*) (1) in optics, a piece of coloured glass or plastic that lets through light of certain wavelengths (p. 119); (2) in radio, an electronic circuit that passes only a certain range of frequencies (p. 119). **filter** (*v*).
friction (*n*) the rubbing of one surface against another. Friction, e.g. with the air, slows down objects that are in motion. The energy of the object's motion is turned into heat.
glow (*v*) to give off light. **glow** (*n*).
graph (*n*) a line drawing that shows how one quantity changes in relation to another, e.g. how the brightness of a star changes with time.
hypothesis (*n*) an unproved idea or theory. **hypothetical** (*adj*).
infinite (*adj*) of something that is without end; too great to be measured or counted. **infinity** (*n*).
integration (*n*) the act of bringing parts together to make a whole, e.g. of rockets, spacecraft, payloads (p. 142) etc. **integrate** (*v*).

interface (*n*) the surface or common boundary where two things meet or join, e.g. the interface between two pieces of equipment, the interface between the atmosphere and space.

jet (*n*) a stream of gas or liquid coming from a hole with force.

latitude (*n*) a coordinate (p. 8) that describes how far north or south an object is. On Earth, latitude is the angle that an object lies north or south of the Earth's equator. In the sky, celestial latitude (p. 12) is measured north or south of the ecliptic (p. 9).

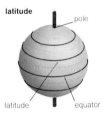

latitude

lava (*n*) rock that has been melted by heat, flowing from a volcano; the same rock when it has cooled again and become solid.

layer (*n*) something that lies at one level, usually part of a series of other layers, e.g. the soil layer.

limit (*n*) the end, edge or border of something; a maximum or minimum value; a set amount beyond which it is not possible or lawful to go.

layer

longitude (*n*) a coordinate (p. 8) that describes how far east or west an object is. On Earth, longitude is measured in degrees east or west of the Greenwich meridian (p. 49). In the sky, celestial longitude (p. 12) is measured along the ecliptic (p. 9) from the vernal equinox (p. 9).

magnetic field the region of space around an object in which the magnetic attraction of that object is felt; lines of force extending into space around a star or planet that control the movement of ionized (p. 116) particles.

maximum (*adj*) the greatest value of something, e.g. maximum brightness.

mean (*adj*) the middle point between two values; average.

mechanical (*adj*) of machines and machinery.

military (*adj*) of armed forces and a nation's defences.

minimum (*adj*) the smallest value of something, e.g. minimum brightness.

momentum (*n*) the product of the mass and velocity of an object. A moving object has momentum.

motion (*n*) the movement of an object; the act of moving.

GENERAL WORDS IN ASTRONOMY · 195

navigate (*v*) to set and guide the course of a ship or any vehicle, especially with respect to a compass reading, a star, etc. **navigation** (*n*).

nominal (*adj*) of something that is planned, intended or aimed for, e.g. a nominal launch (p. 148) date, a nominal trajectory (p. 16).

observe (*v*) to look carefully at something; to watch and to notice, especially from a distance. **observer** (*n*), **observation** (*n*).

orientation (*n*) the position or direction of something, e.g. with respect to the points of the compass.

oscillate (*v*) to swing from side to side or backwards and forwards between two points; to repeat something continuously. **oscillation** (*n*).

parallel (*adj*) of lines that run side by side at the same distance apart, never meeting.

parameter (*n*) a quantity by which things are measured, e.g. length, breadth.

particle (*n*) a very small piece of matter, e.g. a grain of dust or a part of an atom (p. 116).

plane (*n*) a flat surface, real or imaginary.

predict (*v*) to calculate or say what events will happen or where something will lie at a future time, e.g. to predict an eclipse, to predict the position of a planet. **prediction** (*n*).

pressure (*n*) the force of something pressing against another, e.g. the pressure of the atmosphere, or the pressure of a gas in a closed space. Pressure is measured in newtons per square metre (Nm^{-2}) or pascals (Pa). **pressurize** (*v*).

pulsate (*v*) to beat like a heart; to change regularly in size; to give out pulses. **pulsation** (*n*).

pulse (*n*) a short or sudden burst of energy, e.g. a pulse of light or radio waves (p. 120). (*v*) to give out pulses.

radius (*n*) the distance from the centre of a circle or sphere to its edge; half the diameter.

ratio (*n*) the relationship of one quantity to another, e.g. if there is twice as much of quantity A as there is of quantity B, the two quantities are said to be in a ratio of 2 to 1.

record (*v*) to make a note of, in writing etc; to store.

refer (*v*) to make mention of, or look up a note of; to send back or relate to.

oscillate

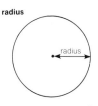

parallel

───────────────

parallel lines

radius

reference (*n*) a place or quantity which other things are measured relative to, e.g. a point of reference; act of referring.

region (*n*) a space in which something lies, or the space near to an object, e.g. an HII region (p. 98).

research (*n*) scientific work to discover new facts. **research** (*v*).

restrict (*v*) to stop or prevent something from going beyond a certain point; to place a boundary or limit on something; to keep within a given area or set of values. **restriction** (*n*).

secular (*adj*) of things that happen over a long period of time.

sensitive (*adj*) of an instrument that is able to notice very small changes or to see very faint objects. **sensitivity** (*n*).

SETI (*abbr*) the search for extraterrestrial intelligence.

signal (*n*) a sign, message or command, e.g. a radio signal. **signal** (*v*).

site (*n*) a place where something happens, e.g. a launch (p. 148) site.

sphere (*n*) a round object shaped like a ball. **spherical** (*adj*).

test (*v*) to try something out; to practise, exercise or examine, e.g. make a test flight. **test** (*n*), (*adj*).

theory (*n*) a set of laws or ideas to explain certain facts and observations. Theories allow events to be predicted.

tilt (*v*) to slope at an angle, e.g. the tilt of a planet's axis. **tilt** (*n*).

tilt

transmit (*v*) to send messages, e.g. by radio; to allow something to pass through, e.g. glass transmits light. **transmission** (*n*).

unit (*n*) a fixed quantity or amount for making measurements, e.g. a unit of length.

velocity (*n*) the speed of an object in a given direction.

visible (*adj*) of something that can be seen. **visibility** (*n*).

volcano (*n*) an opening in the ground from which hot rock and gases come out; the mountain formed by such activity.

volume (*n*) the amount of space that an object takes up.

volcano

APPENDIX ONE

Planets

planet	equatorial diameter (km)	sidereal rotation period	inclination of axis (degrees)	average distance from Sun (million km)	orbital period	inclination of orbit (degrees)	mass (Earth = 1)
Mercury	4,878	58.65 d	0	57.9	87.97 d	7	0.06
Venus	12,104	243.0 d	2	108.2	224.7 d	3.4	0.82
Earth	12,756	23.93 h	23.4	149.6	365.3 d	0	1.0
Mars	6,794	24.62 h	25.2	227.9	687 d	1.9	0.11
Jupiter	142,800	9.84 h	3.1	778.3	11.9 y	1.3	318
Saturn	120,000	10.23 h	26.7	1,427	29.5 y	2.5	95
Uranus	52,000	17.24 h	98	2,870	84.0 y	0.8	14.5
Neptune	48,400	18 h(?)	29	4,496	165 y	1.8	17.2
Pluto	3,000	6.4 d	118	5,900	250 y	17.2	0.002

Moons

planet	satellite	average distance from planet (km)	orbital period (days)	diameter (km)	inclination of orbit to equator of planet (degrees)
Earth	Moon	384,400	27.32	3,476	23.4
Mars	Phobos	9,380	0.32	23	1.0
	Deimos	23,500	1.26	13	2.0
Jupiter	Amalthea	181,300	0.49	240	0.4
	Io	412,600	1.77	3,650	0.0
	Europa	670,900	3.55	3,100	0.5
	Ganymede	1,070,000	7.16	5,200	0.2
	Callisto	1,880,000	16.69	4,800	0.2
	Leda	11,110,000	240	10(?)	26.7
	Himalia	11,470,000	250.6	170	28
	Lysithea	11,710,000	260	20(?)	29
	Elara	11,740,000	260.1	80	28
	Ananke	20,700,000	617	20(?)	147
	Carme	22,350,000	692	25(?)	163
	Pasiphae	23,300,000	735	25(?)	148
	Sinope	23,700,000	758	20(?)	157
Saturn	Mimas	185,000	0.94	400	1.5
	Enceladus	238,000	1.37	500	0.0
	Tethys	295,000	1.89	1,000	1.1
	Dione	377,000	2.74	1,100	0.0
	Rhea	527,000	4.52	1,500	0.3
	Titan	1,222,000	15.95	5,100	0.3
	Hyperion	1,481,000	21.28	300	0.6
	Iapetus	3,560,000	79.33	1,500	14.7
	Phoebe	12,930,000	550.4	150	150.0
Uranus	Miranda	130,000	1.41	480	3.4
	Ariel	191,000	2.52	1,160	0.0
	Umbriel	266,000	4.14	1,190	0.0
	Titania	436,000	8.71	1,610	0.0
	Oberon	583,000	13.46	1,550	0.0
Neptune	Triton	355,000	5.88	3,800	160
	Nereid	5,562,000	359.88	900	28
Pluto	Charon	20,000	6.39	1,200(?)	0 (?)

APPENDIX TWO

Constellations

constellation	popular name	order of size
Andromeda	Andromeda	19
Antlia	air pump	62
Apus	bird of paradise	67
Aquarius	water carrier	10
Aquila	eagle	22
Ara	altar	63
Aries	ram	39
Auriga	charioteer	21
Boötes	herdsman	13
Caelum	chisel	81
Camelopardalis	giraffe	18
Cancer	crab	31
Canes Venatici	hunting dogs	38
Canis Major	greater dog	43
Canis Minor	lesser dog	71
Capricornus	sea goat	40
Carina	keel	34
Cassiopeia	Cassiopeia	25
Centaurus	centaur	9
Cepheus	Cepheus	27
Cetus	whale	4
Chamaeleon	chameleon	79
Circinus	compasses	85
Columba	dove	54
Coma Berenices	Berenice's hair	42
Corona Australis	southern crown	80
Corona Borealis	northern crown	73
Corvus	crow	70
Crater	cup	53
Crux	southern cross	88
Cygnus	swan	16
Delphinus	dolphin	69
Dorado	goldfish	72
Draco	dragon	8
Equuleus	little horse	87
Eridanus	river	6
Fornax	furnace	41
Gemini	twins	30
Grus	crane	45
Hercules	Hercules	5
Horologium	pendulum clock	58
Hydra	water snake	1
Hydrus	lesser water snake	61
Indus	Indian	49
Lacerta	lizard	68
Leo	lion	12
Leo Minor	lesser lion	64
Lepus	hare	51
Libra	scales	29
Lupus	wolf	46
Lynx	lynx	28
Lyra	lyre	52
Mensa	table mountain	75
Microscopium	microscope	66
Monoceros	unicorn	35
Musca	fly	77
Norma	level	74
Octans	octant	50
Ophiuchus	serpent bearer	11
Orion	Orion	26
Pavo	peacock	44
Pegasus	Pegasus	7
Perseus	Perseus	24
Phoenix	phoenix	37
Pictor	painter	59
Pisces	fishes	14
Piscis Austrinus	southern fish	60
Puppis	stern	20
Pyxis	compass	65
Reticulum	net	82
Sagitta	arrow	86
Sagittarius	archer	15
Scorpius	scorpion	33
Sculptor	sculptor	36
Scutum	shield	84
Serpens	serpent	23
Sextans	sextant	47
Taurus	bull	17
Telescopium	telescope	57
Triangulum	triangle	78
Triangulum Australe	southern triangle	83
Tucana	toucan	48
Ursa Major	great bear	3
Ursa Minor	little bear	56
Vela	sails	32
Virgo	virgin	2
Volans	flying fish	76
Vulpecula	fox	55

APPENDIX THREE

Observatories

the numbers in circles at the end of entries refer to map on page 201

Radio Observatories

Algonquin Radio Observatory a radio observatory in Alonquin Park, Ontario, owned by the National Research Council of Canada. Its radio telescope is a dish 46 m in diameter. ①

Arecibo Observatory a radio observatory near Arecibo, Puerto Rico, owned by Cornell University. It has the world's largest single radio astronomy dish, 305 m in diameter, but it is fixed in place and can only point upwards. ②

Effelsberg Radio Observatory the observatory near Bonn of the Max Planck Institute for Radio Astronomy. It has the world's largest radio dish that can be pointed to any part of the sky, 100 m in diameter. ③

Jodrell Bank a radio observatory at Macclesfield, Cheshire, owned by the University of Manchester. Its largest radio telescope is a 76-m dish. ④

Mullard Radio Astronomy Observatory the radio observatory of the University of Cambridge. It has two large aperture synthesis telescopes: the One Mile telescope, consisting of three 18-m dishes in a line 1 mile long; and the Five-Kilometre telescope, consisting of eight 13-m dishes in a line 5 km long. ⑤

National Radio Astronomy Observatory a radio observatory at Green Bank, West Virginia, owned by a group of American universities. Its largest radio telescopes are dishes of 91-m and 43-m diameter. The National Radio Astronomy Observatory also runs the Very Large Array. ⑥

Parkes Observatory Australia's national radio astronomy observatory, near Parkes, New South Wales, owned by the Commonwealth Scientific and Industrial Research Organisation (CSIRO). Its largest radio telescope is a 64-m dish. ⑦

Very Large Array an aperture synthesis radio telescope near Socorro, New Mexico, run by the National Radio Astronomy Observatory. It consists of 27 dishes each 25 m in diameter arranged in a Y-shape with arms up to 21 km long. ⑧

Westerbork Radio Observatory an aperture synthesis radio telescope near Groningen, Netherlands, owned by the Netherlands Foundation for Radio Astronomy. It has twelve 25-m dishes in a line 1.5 km long. ⑨

Optical Observatories

Anglo-Australian Observatory an observatory at Siding Spring, New South Wales, owned by the UK and Australia. Its largest telescope is a 3.9-m reflector. ⑩

Cerro Tololo Inter-American Observatory an observatory near La Serena in Chile, owned by a group of US universities. Its largest telescope is a 4-m reflector. ⑪

European Southern Observatory an observatory at La Silla in Chile, owned by a group of European countries. Its largest telescope is a 3.6-m reflector. ⑫

Kitt Peak National Observatory an observatory near Tucson, Arizona, owned by a group of American universities. Its largest telescope is a 4-m reflector, as well as the world's largest solar telescope with a 1.5-m mirror. ⑬

Lick Observatory an observatory on Mount Hamilton, California, owned by the University of California. Its largest telescopes are a 3-m reflector and the 91-cm Lick refractor. ⑭

Mauna Kea Observatory an observatory on Mauna Kea in Hawaii, owned by the University of Hawaii. On Mauna Kea are the 3.7-m Canada-France-Hawaii reflector and the 3.8-m United Kingdom Infrared Telescope. ⑮

McDonald Observatory an observatory near Fort Davis, Texas, owned by the University of Texas. Its largest telescope is a 2.7-m reflector. ⑯

Mount Wilson Observatory an observatory near Los Angeles, California. Its largest telescope is the 2.5-m Hooker reflector. ⑰

Palomar Observatory an observatory near Pasadena, California. Its largest telescope is the 5-m Hale reflector, plus a 1.2-m Schmidt telescope. ⑱

Roque de los Muchachos Observatory an observatory on La Palma in the Canary Islands, owned by a group of European countries. Its largest telescopes are the 2.5-m Isaac Newton reflector and the 4.2-m William Herschel reflector, both provided by the United Kingdom. ⑲

Royal Greenwich Observatory the main United Kingdom observatory, at Herstmonceux in Sussex. It provides the Greenwich time service. Its largest telescopes are at the Roque de los Muchachos Observatory. ⑳

Siding Spring Observatory an observatory at Siding Spring, New South Wales. At Siding Spring is the 1.2-m United Kingdom Schmidt telescope. Next to it is the Anglo-Australian Observatory. ㉑

Smithsonian Astrophysical Observatory an observatory in Cambridge, Massachusetts. Its largest telescope is the Multiple Mirror Telescope on Mount Hopkins, near Tucson, Arizona. This telescope consists of six mirrors 1.8 m in diameter, which are arranged to give the same view of the sky as a single mirror of 4.5-m aperture. ㉒

US Naval Observatory an observatory owned by the US government in Washington, D.C. Its main telescope is a 66-cm refractor. ㉓

Yerkes Observatory an observatory at Williams Bay, Wisconsin, owned by the University of Chicago. Its largest telescope is a 1-m refractor, the largest refracting telescope in the world. ㉔

Zelenchukskaya Observatory an observatory in the Caucasus mountains of the Soviet Union, owned by the Soviet Academy of Science. It contains the world's largest optical telescope, a 6-m reflector. Another major instrument there is the RATAN 600 radio telescope, built in a circle 600 m in diameter. ㉕

APPENDIX FOUR

Space Centres

the numbers in circles at the end of entries refer to map on page 201.

- **National Aeronautics and Space Administration** NASA. A branch of the United States government, set up in 1958 to explore space for peaceful purposes. Its head office is in Washington, D.C. ㉖
- **Kennedy Space Centre** KSC. A part of NASA; the place from where most of NASA's space missions are launched. All NASA manned space flights, all NASA space probes, and most NASA Earth satellites have been launched from the Kennedy Space Centre. The Kennedy Space Centre is at Cape Canaveral, Florida. ㉗
- **Cape Canaveral** the place in Florida where the Kennedy Space Centre is. Cape Canaveral is on the shore of the Atlantic Ocean. From 1963 to 1973 it was known as Cape Kennedy. Not only NASA uses Cape Canaveral. The U.S. Air Force launches its own rockets and satellites from the Cape Canaveral Air Force Station, which is next to the Kennedy Space Centre. ㉘
- **Eastern Test Range** ETR. The rocket launching and tracking range, run by the U.S. Air Force, that stretches from Cape Canaveral southeastwards over the Atlantic Ocean. ㉙
- **Johnson Space Centre** JSC. A part of NASA at Houston, Texas, the home of mission control for manned space flights. It is NASA's main centre for the planning of Space Shuttle missions and also for the training of astronauts. ㉚
- **Jet Propulsion Laboratory** JPL. A part of NASA at Pasadena, California, run by the California Institute of Technology. JPL is the centre of NASA's Deep Space Network of tracking stations, and controls the missions of space probes e.g. Viking and Voyager. ㉛
- **Marshall Space Flight Centre** MSFC. A part of NASA at Huntsville, Alabama, in charge of the planning and building of rocket engines such as those in the Saturn rockets and the Space Shuttle. It is also concerned with major Space Shuttle payloads, e.g. Spacelab and the Hubble Space Telescope. ㉜
- **Goddard Space Flight Centre** GSFC. The centre of NASA's worldwide satellite tracking and communications systems, at Greenbelt, Maryland. Readings from NASA unmanned Earth satellites are received and stored at GSFC. ㉝
- **Vandenberg Air Force Base** VAFB. A launching site at Lompoc, California, used by the U.S. Air Force and NASA. Satellites are put into polar orbit from Vandenberg, many of them for military purposes. Vandenberg Air Force Base is a launch and landing site for flights of the Space Shuttle in polar orbit. Missiles are also test fired from Vandenberg. ㉞

APPENDIX FOUR/SPACE CENTRES · **203**

Western Test Range WTR. The rocket launching and tracking range, run by the U.S. Air Force, that stretches southwards over the Pacific Ocean from the Vandenberg Air Force Base. ㉟

Wallops Island an island in Virginia, USA, on the coast of the Atlantic Ocean, from where NASA launches rockets and satellites. Most of the launches from Wallops Island have been of sounding rockets, but a number of small satellites for scientific studies have been put into orbit from Wallops Island by Scout rockets. ㊱

Edwards Air Force Base a landing site for the Space Shuttle, in the Mojave Desert of California. The first Space Shuttle to return from orbit landed at Edwards Air Force Base, and it is the main back-up landing site when the weather is too bad for the Shuttle to land at the Kennedy Space Centre. ㊲

Dryden Flight Research Facility a part of NASA, at the Edwards Air Force Base, California, concerned with the flight of aircraft at great height and great speed, including re-entry and landing of spacecraft such as the Space Shuttle. ㊳

White Sands a missile firing range at Las Cruces, New Mexico, from where many early sounding rockets were launched. White Sands is a back-up landing site for the Space Shuttle. The ground station for the Tracking and Data Relay Satellites is placed there. ㊴

Ames Research Centre ARC. A part of NASA, in charge of the Pioneer series of space probes. The Ames Research Centre is also concerned with the effects of spaceflight on the human body, and the search for life off Earth. ㊵

Tyuratam the main Soviet launch site, northeast of the Aral Sea. Soviet manned spaceflights and space probes are all launched from here, as was Sputnik 1 and many later satellites. ㊶

Baikonur the name given by the Russians to their main Soviet launch centre at Tyuratam. Baikonur is actually a town about 300 km to the northeast of Tyuratam. ㊷

Plesetsk a Soviet launch site 170 km south of Archangel in northern Russia, used for sending satellites into polar orbit. Many of the launches from Plesetsk are for military purposes. ㊸

Kapustin Yar a small Soviet launch site about 100 km southeast of Volgograd, used for launching sounding rockets and for putting small satellites into orbit. ㊹

European Space Agency ESA. A group of 11 European nations, formed in 1975 to explore space for peaceful purposes. Its head office is in Paris. The members of the European Space Agency are: Belgium, Denmark, France, Germany, Ireland, Italy, the Netherlands, Spain, Sweden, Switzerland and the United Kingdom. ㊺

Kouru the place in French Guiana, on the Atlantic coast of South America, from where the European Space Agency launches its Ariane rocket. Its official name is the Guiana Space Centre. ㊻

APPENDIX FIVE

Great names in astronomy and astronautics

Adams, John Couch (1819-1892) English astronomer, predicted that there was a planet beyond Uranus, which was causing perturbations in the motion of Uranus. Similar calculations were made in France by Urbain Leverrier. The planet is now known at Neptune.

Aldrin, Edwin "Buzz" (b. 1930) American astronaut, landed on the Moon with Neil Armstrong during the Apollo 11 flight in July 1969.

Aristarchus (c. 275 BC) Greek scientist, first person to put forward the idea that the Sun, not the Earth, was the centre of the solar system.

Aristotle (384-322 BC) Greek scientist, said that the Earth was round, but did not think that it rotated or went around the Sun.

Armstrong, Neil Alden (b. 1930) American astronaut, was the first person to walk on the Moon. He landed on the Moon with Edwin Aldrin during the Apollo 11 flight. As he stepped onto the lunar surface on July 21, 1969, he said: "That's one small step for a man; one giant leap for mankind."

Baade, Walter (1893-1960) German-born American astronomer, discovered that there are two populations of stars, Population I and Population II.

Bayer, Johann (1572-1625) German astronomer, made the first map of the whole sky, on which he introduced the system of Bayer letters.

Bessel, Friedrich Wilhelm (1784-1846) German astronomer, made the first measurement of the parallax of a star.

Bethe, Hans Albrecht (b. 1906) German-born American physicist, showed that stars get their energy from nuclear reactions.

Bradley, James (1693-1762) English astronomer, discovered the aberration of starlight and nutation.

Brahe, Tycho (1546-1601) Danish astronomer, made exact observations of the planets from which Johannes Kepler worked out his laws of planetary motion.

von Braun, Wernher (1912-1977) German-born American rocket engineer, built the V2 missile in Germany during World War II, then went to the United States where he was head of the team that built the Saturn IB and Saturn V rockets.

Cannon, Annie Jump (1863-1941) American astronomer, started the system of grouping stars into spectral types.

Cassini, Jean Dominique (1625-1712) French astronomer, studied the planets, discovered four moons of Saturn and the Cassini division in the rings of Saturn.

Chandrasekhar, Subrahmanyan (b. 1910) Indian-born American astrophysicist, calculated that a white dwarf star cannot have a mass greater than 1.4 times that of the Sun, a value known as Chandrasekhar's limit. In 1983 he was awarded the Nobel prize for physics.

Clarke, Arthur C. (b. 1917) English writer, put forward the idea of satellites for communication in geostationary orbit, now sometimes known as Clarke orbit.

Collins, Michael (b. 1930) American astronaut, remained in the command module of Apollo 11 while Neil Armstrong and Edwin Aldrin landed on the surface of the Moon in July 1969.

Copernicus, Nicolaus (1473-1543) Polish astronomer, in 1543 put forward the heliocentric system, which said that the Earth is not the centre of the Universe as thought until then, but that it is an ordinary planet orbiting the Sun.

Doppler, Christian Johann (1803-1853) Austrian physicist, discovered the Doppler effect.

Eddington, Arthur Stanley (1882-1944) English astrophysicist, calculated that the insides of stars must be in the form of an ionized gas at a temperature of many millions of degrees, and discovered the mass-luminosity relation for stars.

Einstein, Albert (1879-1955) German physicist, one of the greatest scientists of all time, put forward the theory of relativity. He won the Nobel prize for physics in 1921.

Eratosthenes (c. 240 BC) Greek astronomer, the first person to measure the size of the Earth correctly.

Flamsteed, John (1646-1719) the first Astronomer Royal of England, made an important list of the positions of nearly 3,000 stars, each of which were given Flamsteed numbers.

Fraunhofer, Joseph (1787-1826) German physicist, discovered the dark Fraunhofer lines that cross the spectrum of the Sun.

Gagarin, Yuri (1934-1968) Soviet cosmonaut, the first person to travel in space. He made one orbit of the Earth on April 12, 1961 in Vostok 1.

Galileo Galilei (1564-1642) Italian scientist, the first person to study the sky with a telescope. He discovered four moons of Jupiter, the craters on the Moon, and the phases of the planet Venus. His observations helped prove the heliocentric system of Copernicus.

Galle, Johann Gottfried (1812-1910) German astronomer, discovered the planet Neptune on September 23, 1846, after its position had been predicted by Urbain Leverrier and J.C. Adams.

Glenn, John Herschel (b. 1921) American astronaut, orbited the Earth three times in a Mercury spacecraft on February 20, 1962. He was the first American to orbit the Earth; two earlier Mercury flights by American astronauts were both sub-orbital.

Goddard, Robert Hutchings (1882-1945) American engineer, built and flew the world's first liquid-propellant rocket on March 16, 1926.

Grissom, Virgil "Gus" (1926-1967) American astronaut, the first person to go into space twice. He made a sub-orbital flight in a Mercury spacecraft on July 21, 1961, and orbited the Earth three times in a Gemini spacecraft on March 23, 1965. He died when a fire broke out in the spacecraft during a practice countdown for the first Apollo flight.

Hale, George Ellery (1868-1938) American astronomer, invented the spectroheliograph; discovered that sunspots were cooler areas on the Sun with strong magnetic fields. He set up the 1-m refracting telescope at Yerkes Observatory, the 2.5 m reflector at Mount Wilson Observatory, and the 5-m reflector at Palomar Observatory.

Halley, Edmond (1656-1742) English astronomer, calculated the orbit of the comet that is now known as Halley's comet. He also discovered that stars have proper motions.

Herschel, William (1738-1822) German-born English astronomer, discovered the planet Uranus on March 13, 1781. He discovered that many stars close together in the sky are in orbit around each other, i.e. they are double stars. He also showed that the stars of the Galaxy are arranged in a flat disc shape.

Herschel, John (1792-1871) English astronomer, son of William Herschel, continued his father's work of examining the sky, discovering star clusters, nebulae and galaxies. The discoveries of the Herschels gave rise to the NGC catalogue.

Hertzsprung, Ejnar (1873-1967) Danish astronomer, found that stars are divided into two main groups, giants and dwarfs. He was the first to draw up what is now known as the Hertzsprung-Russell diagram.

Hevelius, Johannes (1611-1687) Polish astronomer, made maps of the Moon and stars. Seven new constellations were first shown on the star map of Hevelius.

Hipparchus (c. 140 BC) Greek astronomer, discovered the effect of precession and was the first person to list stars according to their brightness or magnitude.

Hubble, Edwin Powell (1889-1953) American astronomer, discovered the expansion of the Universe. He proved that our Milky Way is not the only galaxy in space, and he started the Hubble classification of galaxies.

Huygens, Christiaan (1629-1695) Dutch scientist, put forward the idea that light travels as a wave. He observed the planets, discovering the first markings on Mars and Saturn's largest moon, Titan. He explained the nature of Saturn's rings.

Jansky, Karl Guthe (1905-1950) American radio engineer, the first person to pick up radio waves from our Galaxy. His discovery started the study of radio astronomy.

Kepler, Johannes (1571-1630) German mathematician and astronomer, discovered the three laws governing the movement of planets in their orbits, known as Kepler's laws. His work finally proved the heliocentric system of Copernicus that the Earth is a planet orbiting the Sun.

Lacaille, Nicolas Louis de (1713-1762) French astronomer, mapped the stars of the southern hemisphere from the Cape of Good Hope, and made up 14 new constellations in the southern skies.

Lagrange, Joseph Louis (1736-1813) French mathematician, studied the orbits of the planets and calculated the Lagrangian points.

Laplace, Pierre Simon (1749-1827) French mathematician and astronomer, put forward the theory that the solar system formed from a cloud of gas around the Sun, similar to present-day ideas on the subject.

Leavitt, Henrietta Swan (1868-1921) American astronomer, discovered the period-luminosity relation of Cepheid variable stars.

Lemaître, Georges Edouard (1894-1966) Belgian astronomer, put forward the Big Bang theory of the origin of the Universe.

Leonov, Alexei (b. 1934) Soviet cosmonaut, the first man to "walk" in space. He went outside his Voskhod 2 spacecraft for 10 minutes on March 18, 1965.

Leverrier, Urbain Jean Joseph (1811-1877) French mathematician, predicted that an unseen planet lay beyond Uranus from the perturbations in the movement of Uranus. He calculated the new planet's position, and it was discovered by J.G. Galle.

Lowell, Percival (1855-1916) American astronomer, thought there were canals on Mars. He set up his own observatory in Arizona to study the planets. He started the search that led to the discovery of Pluto.

Messier, Charles (1730-1817) French astronomer, discovered over 15 comets and made a famous list of over 100 bright nebulae, galaxies and star clusters. The objects on Messier's list are still known to astronomers by their M (for Messier) numbers.

Newton, Isaac (1642-1727) English scientist, one of the greatest scientists of all time, worked out the laws of motion of bodies and the law of gravity. He also designed the form of telescope known as the Newtonian reflector.

Oort, Jan Hendrik (b. 1900) Dutch astronomer, studied the spiral arms of our Galaxy by radio astronomy. He put forward the idea that comets come from a cloud around our solar system, the Oort cloud.

Piazzi, Giuseppe (1746-1826) Italian astronomer, discovered Ceres, the first asteroid, on January 1, 1801.

Ptolemy (c. 100-c. 178) Greek scientist, taught that the Earth was the centre of the Universe and that everything else went around it. This was known as the Ptolemaic system, and it was widely believed until the time of Copernicus.

Russell, Henry Norris (1877-1957) American astronomer, studied stellar evolution and eclipsing binary stars. He found that a star's brightness is related to its colour, as shown on the Hertzsprung-Russell diagram.

Shapley, Harlow (1885-1972) American astronomer, discovered in 1918 that the Sun is not at the centre of the Galaxy, as thought until then, but that it is about two-thirds of the way to the edge. He made his discovery as a result of measuring the distances to globular clusters.

Shepard, Alan Bartlett (b.1923) American astronaut, the first American to go into space. He made a sub-orbital flight in a Mercury spacecraft on May 5, 1961. He also landed on the Moon in February 1971 during the Apollo 14 mission.

Struve, Friedrich George Wilhelm (1793-1864) German astronomer, noted for his observations of double stars of which he discovered over 1,000.

Tereshkova, Valentina (b. 1937) Soviet cosmonaut, the first woman to fly in space. She orbited the Earth for three days in June 1963 in the spacecraft Vostok 6.

Titov, Gherman (b. 1935) Soviet cosmonaut, the first person to spend a complete day in space. He made 17 orbits of the Earth in Vostok 2 in August 1961.

Tombaugh, Clyde William (b. 1906) American astronomer, discovered the planet Pluto on February 18, 1930.

Tsiolkovsky, Konstantin (1857-1935) Russian mathematician, showed that rockets, satellites and space stations were possible before the first aeroplanes had flown. He pointed out that rockets needed to be made in stages to escape from the Earth, and put forward the idea of using liquid hydrogen and liquid oxygen as propellants; these are actually the propellants used in the engines of the Space Shuttle.

White, Edward Higgins (1930-1967) American astronaut, the first American to "walk" in space; he left his Gemini spacecraft for 20 minutes on June 3, 1965. He died in a fire during a practice countdown for the first Apollo flight.

Young, John Watts (b. 1930) American astronaut, landed on the Moon with Apollo 16 in April 1972 and was commander of the first Space Shuttle flight in April 1981. Young also made two Gemini spaceflights, the Apollo 10 flight that practised for the first Moon landing, and he commanded the Space Shuttle flight that carried the first Spacelab.

APPENDIX SIX

Space abbreviations and acronyms

AOS acquisition of signal (from a spacecraft).

APT automatic picture transmission (from weather satellites).

APU auxiliary power unit, a device on the Space Shuttle that provides power to gimbal the main engines.

ARC Ames Research Centre.

ASAT anti-satellite, i.e. a missile or other weapon to shoot down a satellite.

ASTP Apollo-Soyuz Test Project, the joint mission in which an American Apollo docked in orbit with a Soviet Soyuz in July 1975.

ATS Applications Technology Satellite, a series of six NASA satellites that tried out new equipment for communications and weather satellites in geostationary orbit.

CAPCOM capsule communicator, an astronaut at mission control who talks to astronauts aboard a spacecraft.

CM command module.

CSM command and service modules (of the Apollo spacecraft).

DSN Deep Space Network.

EAFB Edwards Air Force Base.

ECS (1) European Communications Satellite; (2) environmental control system, i.e. to control the conditions inside a manned spacecraft.

EI entry interface.

EMU extravehicular mobility unit.

ESA European Space Agency.

ESSA a series of weather satellites owned by the Environmental Science Services Administration of the United States.

ET (1) External Tank; (2) extraterrestrial, i.e. not of this Earth.

ETR Eastern Test Range.

EURECA European retrievable carrier, a frame to which equipment could be fixed and left in orbit, to be picked up again later by the Space Shuttle.

EVA extravehicular activity.

FOBS fractional orbit bombardment system, a Soviet space weapon.

GAS getaway special.

GEO geostationary Earth orbit.

GLOW gross lift-off weight, i.e. the weight of a rocket at launch including fuel, payload, etc.

GOES Geostationary Operational Environmental Satellite, a series of American weather satellites in geostationary orbit.

GPS global positioning system, a system of navigation using Navstar satellites.

GSFC Goddard Space Flight Centre.

GTO geostationary transfer orbit.

HAC heading alignment cylinder.

HEAO High Energy Astronomy Observatory.

HST Hubble Space Telescope.

ICBM intercontinental ballistic missile.

IMU inertial measurement unit.

IRAS Infra Red Astronomy Satellite.

IRBM intermediate range ballistic missile.

ISEE International Sun-Earth Explorer.

IUE International Ultraviolet Explorer.

IUS Inertial Upper Stage.

JPL Jet Propulsion Laboratory.

JSC Johnson Space Centre.

KSC Kennedy Space Centre.

LDEF Long Duration Exposure Facility, a large satellite to which instruments and experiments are fixed; it is left in orbit by the Space Shuttle and brought back to Earth after several months.

LEO low Earth orbit, e.g. the height reached by the Space Shuttle.

LM lunar module.

LOS loss of signal (from a spacecraft).

LOX liquid oxygen, the usual oxidizer for rockets.

LRV lunar roving vehicle.

MCC mission control centre.

MECO main engine cut-off (of a rocket during launch).

MET mission elapsed time, i.e. time since a spacecraft was launched.

MMS multi-mission modular spacecraft.

MMU manned manoeuvring unit.

MSFC Marshall Space Flight Centre.

NASA National Aeronautics and Space Administration.

NASCOM NASA communications network, a worldwide system of communications, centred on the Goddard Space Flight Centre.

NERVA nuclear engine for rocket vehicle application, a nuclear rocket that NASA tried to build in the 1960s but which was never completed.

NOAA a series of weather satellites owned by the National Oceanic and Atmospheric Administration of the United States.

OAO Orbiting Astronomical Observatory.

OMS orbital manoeuvring system.

OSO Orbiting Solar Observatory.

PAM Payload Assist Module.

PLBD payload bay doors (of the Space Shuttle).

PLSS portable life-support system.

RCS reaction control system.

RFNA red fuming nitric acid, an oxidizer used in liquid-fuel rockets.

RMS remote manipulator system.

S-IB the first stage of a Saturn IB rocket.

S-IC the first stage of a Saturn V rocket.

S-II the second stage of a Saturn V rocket.

S-IVB the second stage of a Saturn IB rocket, and the third stage of a Saturn V rocket.

SAS Small Astronomy Satellite.

SM service module.

SMM Solar Maximum Mission.

SPS (1) service propulsion system; (2) solar power satellite.

SRB Solid Rocket Booster.

SSME Space Shuttle main engine.

SSUS spinning solid upper state = Payload Assist Module.

STDN Space Tracking and Data Network.

STS Space Transportation System.

TACAN tactical air navigation.

TAEM terminal area energy management.

TDRS Tracking and Data Relay Satellite.

TIG time of ignition, i.e. of a rocket engine.

UDMH unsymmetrical dimethyl hydrazine, a liquid fuel for rockets.

VAB Vehicle Assembly Building, a large building at Kennedy Space Centre where rockets are prepared for launch.

VAFB Vandenberg Air Force Base.

WI Wallops Island.

WTR Western Test Range.

APPENDIX SEVEN

Symbols used in astronomy

LETTER/SYMBOL	MEANING/QUANTITY
a	major axis
b	galactic latitude
e	eccentricity
g	acceleration due to gravity
i	inclination
l	galactic longitude
n	mean motion
P	period
Q	aphelion
q	dynamic pressure
q	perihelion
T	time of perihelion passage
α	right ascension
β	celestial latitude
δ	declination
λ	celestial longitude
λ	wavelength
μ	proper motion
π	parallax
Ω	longitude of the ascending node
ω	argument of perihelion
$\bar{\omega}$	longitude of perihelion
Å	ångström (10^{-10}m)
H_0	Hubble's constant
$1/H_0$	Hubble time
q_0	deceleration parameter

APPENDIX EIGHT

International System of Units (SI)

PREFIXES

PREFIX	FACTOR	SIGN	PREFIX	FACTOR	SIGN
milli	$\times 10^{-3}$	m	kilo	$\times 10^{3}$	k
micro	$\times 10^{-6}$	μ	mega	$\times 10^{6}$	M
nano	$\times 10^{-9}$	n	giga	$\times 10^{9}$	G
pico	$\times 10^{-12}$	p	tera	$\times 10^{12}$	T

BASE UNITS

UNIT	SYMBOL	MEASUREMENT
metre	m	length
kilogram	kg	mass
second	s	time
ampere	A	electric current
kelvin	K	temperature
mole	mol	amount of substance
candela	cd	luminous intensity

COMMON DERIVED UNITS

UNIT	SYMBOL	MEASUREMENT
newton	N	force
joule	J	energy, work
hertz	Hz	frequency
pascal	Pa	pressure
coulomb	C	quantity of electric charge
volt	V	electrical potential
ohm	Ω	electrical resistance

Index

aberration 125
aberration of starlight 89
ablation 66
abort 150
absolute 192
absolute magnitude 68
absolute zero 118
absorb 119
absorption lines 121
absorption nebula 98
accelerate 150
acceleration 150
accelerometer 155
accretion 27
accretion disk 87
Achernar 192
achondrite 67
achromatic 125
acquisition 155
acronym 192
active prominence 32
ADS 97
aerial 135
aerolite 66
Agena 176
AGK 196
airglow 46
airlock 163
Airy disc 126
albedo 27
Aldebaran 92
Aldrin, Edwin 204
Algol 85
almanac 97
Alpha Centauri 92
Alpha Crucis 92
Altair 92
altazimuth mounting 131
altitude^1 11
altitude^2 153
Amor asteroid 64
amplifier 135

amplitude 81
analemma 50
Andromeda galaxy 105
ångström 119
angular diameter 15
angular distance 15
angular measure 15
angular momentum 26
annular eclipse 41
anomalistic month 55
anomalistic year 54
anomaly 21
ansa 59
Antares 93
antenna 135
ap-, apo- 18
apastron 18
aperture 123
aperture synthesis 136
aphelion 18
aplanatic 126
apoapsis see apsis 18
apochromat 125
apogee 18
apogee kick motor 151
Apollo 164
Apollo asteroid 64
apparent 192
apparent magnitude 68
apparent solar time 49
apparition 15
appulse 42
apsidal motion 18
apsis 18
arc 15
archaeoastronomy 139
arc minute 15
arc second 15
Arcturus 93
area ratio 147
argument of perihelion 20
Ariane 174

Ariel 189
armillary sphere 140
array 135
artificial satellite 152
ascending node
   see node 20
ascent 149
ashen light 57
aspect 25
asterism 95
asteroid 64
asteroid belt 64
astigmatism 126
astrodynamics 16
astrograph 130
astrolabe 140
astrology 140
astrometric binary 87
astrometry 89
astronaut 162
astronautics 152
Astronomical
   Almanac 97
astronomical unit 24
astronomy 192
astrophysics 112
Atlantis 174
Atlas 174
atmosphere 43
atmospheric
   extinction 47
atom 116
atomic clock 52
atomic time 52
attitude 156
attraction 192
A-type 176
aurora 45
aurora australis 45
aurora borealis 45
auroral oval 45
autumnal equinox 9
average 192

INDEX · 215

avionics 155
axis 192
azimuth 11

backup 142
Baily's beads 42
ballistic 141
ballistic missile 141
Barnard's star 93
barred spiral
  galaxy 104
barycentre 22
baseline 136
Bayer letter 96
beamwidth 135
Besselian year 54
Beta Canis Majoris
  stars 82
Beta Centauri 93
Beta Cephei stars 82
Betelgeuse 93
Big Bang 110
binary pulsar 78
binary star 85
bi-propellant 144
black body 122
black hole 79
blink microscope 134
BL Lacertae
  object 107
blue giant 72
blueshift 115
Bode's law 27
boilerplate 152
Bok globule 99
bolide 62
bolometer 133
bolometric
  correction 69
bolometric
  magnitude 69
Bonner
  Durchmusterung 96
booster 142
Boss General
  Catalogue 97
boundary 192
bremsstrahlung 117
bright nebula 98
B-type 176

burn 147
burnout 147
butterfly diagram 30

calendar 52
Canals of Mars 57
Canopus 93
Capella 94
capsule 162
captured 19
captured rotation 26
carbonaceous
  chondrite 67
carbon cycle 70
carbon-nitrogen-
  oxygen cycle 70
Cassegrain
  reflector 129
Cassini's division 58
Cassiopeia A 77
cataclysmic variable 83
catadioptric
  telescope 129
cD galaxy 104
celestial 192
celestial equator 8
celestial latitude 12
celestial longitude 12
celestial mechanics 16
celestial pole 9
celestial sphere 8
Centaur 176
Centaurus A 107
central peak 34
centre of mass 22
Cepheid instability
  strip 82
Cepheid variable 81
Ceres 65
Challenger 174
chamber pressure 146
Chandrasekhar limit 76
charge-coupled
  device 133
chemical rocket 143
chondrite 67
chondrule 67
chromatic
  aberration 125
chromosphere 31

circumpolar 11
Clarke orbit 23
classify 192
closed Universe 109
cluster 192
cluster of galaxies 106
cluster variable 82
Coalsack 99
coelostat 130
collimation 126
colour index 74
colour-magnitude
  diagram 72
Columbia 173
colure 11
coma$_1$ 126
coma^2 60
Coma cluster 106
combustion 146
combustion
  chamber 146
comet 60
comet family 61
command module 165
commensurable 22
communicate 192
communications
  satellite 179
comparator 134
Compton effect 118
concave 123
cone 192
conjunction 25
constant 192
constellation 95
consumable 163
contact binary 87
contingency 192
continuous
  creation 111
continuous
  spectrum 121
continuum 121
contract 192
control 192
converge 192
convergent-divergent
  nozzle 147
convergent point 88
convex 123

coordinate 8
Copernican system 139
Copernicus
 satellite 188
Cordoba
 Durchmusterung 97
core segment 172
corona 32
coronagraph 131
coronal hole 32
corrector plate 129
COS-B 190
cosmic 192
cosmic abundance of
 elements 117
cosmic background
 radiation 110
cosmic rays 117
cosmodrome 148
cosmogony 108
cosmological 108
cosmological
 principle 108
cosmology 108
cosmonaut 162
Cosmos 178
coudé focus 124
countdown 149
Crab nebula 77
Crab pulsar 78
crater 34
crepe ring *see* rings of
 Saturn 58
crescent 34
crew 162
crossrange 161
crust 192
cryogenic 144
C-type 176
culmination 11
cusp 34
cutoff 147
cycle 192
Cygnus A 107
Cygnus X-1 79
cynthion 33
Cytherean 57

Dall-Kirkham
 reflector 129

dark nebula 98
data 192
Dawes limit 125
day 48
decay 159
decelerate 160
deceleration 160
deceleration
 parameter 109
declination 8
declination axis 132
decoupling 111
deep space 154
Deep Space
 Network 154
deferent 139
degenerate matter 76
degree 15
de Laval nozzle 147
Delta 174
Delta Cephei 81
Delta Scuti stars 82
delta-V 157
dense 192
density 192
density wave 104
deorbit 159
deploy 151
descending node *see*
 node 20
de-spun 158
destruct 150
detail 193
diagram 193
diameter 193
diamond ring 42
dichotomy 57
differential rotation 26
diffraction 120
diffraction grating 127
diffraction ring 126
diffuse nebula 99
dim 193
direct ascent 23
direct broadcast
 satellite 180
direct motion 14
dirty snowball 60
disc 26
Discovery 174

disc population 101
dispersion 127
distance modulus 91
diurnal 48
diurnal parallax 13
diverge 193
docking 157
dome^1 35
dome^2 137
Doppler effect 114
double-line spectro-
 scopic binary 87
double star 85
doublet 125
downlink 155
downrange 151
draconic month 55
drag 156
D-type 176
dust tail 60
dwarf Cepheid 82
dwarf nova 84
dwarf star 72
dynamical time 51
dynamic pressure 150

early-type star 73
Earth 43
Earth-crosser 64
Earth-grazer 64
earthshine 34
eccentricity 18
eclipse 38
eclipse season 39
eclipse year 39
eclipsing binary 85
ecliptic 9
ecliptic limit 39
effective
 temperature 122
Einstein
 observatory 189
ejecta 35
electric rocket 143
electromagnetic
 radiation 118
electron 116
electron
 degeneracy 76
element^1 116

element^2 125
elements of an orbit 19
ellipse 16
elliptical galaxy 104
elongation 25
emersion 42
emission lines 121
emission nebula 98
emission spectrum 121
emit 119
Encke's comet 61
Encke's division 59
energy 193
Enterprise 173
entry interface 160
ephemeris 21
ephemeris time 51
epicycle 139
epoch 21
equation 193
equation of the
 centre 36
equation of time 50
equatorial bulge 26
equatorial horizontal
 parallax 12
equatorial
 mounting 131
equinoctial colure see
 colure 11
equinox 9
ergosphere 79
Eros 65
eruptive variable 83
escape velocity 19
evection 36
event horizon 79
exhaust 147
exhaust velocity 147
exobiology 193
Exosat 191
exosphere 44
expand 193
expansion of the
 Universe 108
expansion ratio 146
experiment 193
experiment
 segment 172
explore 193

Explorer 178
explosive bolt 151
extend 193
External Tank 170
extinction 69
extragalactic 103
extraterrestrial 193
extravehicular
 activity 163
extravehicular mobility
 unit 164
eyepiece 123

facula 30
fairing 142
fall 66
F corona 32
ferret satellite 183
field of view 124
field star 89
filament 32
filter 193
find 66
fireball 62
first contact 41
first point of Aries 9
first quarter 34
FK 96
flame deflector 149
Flamsteed number 96
flare^1 31
flare^2 162
flare star 83
flash spectrum 31
flat 123
flight deck 171
flocculus 31
flyby 23
focal length 123
focal plane 124
focal ratio 124
focus^1 16
focus^2 123
forbidden lines 122
fourth contact 41
Fraunhofer lines 121
free fall 152
frequency 119
friction 193
F-type 177

fuel 143
fuel cell 158
full Moon 33
fundamental
 catalogue 96
fundamental star 96

g 150
galactic centre 102
galactic cluster 88
galactic
 coordinates 102
galactic disc 101
galactic equator 102
galactic halo 101
galactic latitude 102
galactic longitude 103
galactic nucleus 101
galactic plane 101
galactic pole 103
galactic rotation 102
galactic year 102
galaxy 104
Galilean satellite 58
Galileo 187
gamma rays 120
gas tail 60
gegenschein 46
Gemini 166
general theory of
 relativity 113
geo- 43
geocentric
 coordinates 12
geocentric parallax 13
geocentric system 138
geodesy 43
geoid 43
geostationary orbit 23
geosynchronous
 orbit 22
getaway special 173
g force 150
giant planet 24
giant star 72
gibbous 34
gimbal^1 147
gimbal^2 147
Giotto 187
glideslope 162

218 · INDEX

globular cluster 88
globule 99
glow 193
gnomon 56
Gould's belt 103
grain 145
granulation 29
graph 193
gravitation 112
gravitational collapse 112
gravitational field 112
gravitational lens 114
gravitational redshift 115
gravitational wave 112
gravity 112
grazing occultation 42
great circle 10
greatest elongation 25
greenhouse effect 57
Greenwich Mean Time 49
Greenwich meridian 49
Gregorian calendar 53
Gregorian reflector 129
ground track 154
G-type 177
Gum nebula 100
gyroscope 155

HI region 98
HII region 98
hadron era 111
Halley's comet 61
halo population 101
hatch 163
Hayashi track 72
head 60
heading alignment cylinder 161
heat shield 160
heavy element 116
heli, helio- 28
heliacal rising 15
heliacal setting 15
heliocentric 28
heliocentric coordinates 13
heliocentric system 138

heliometer 131
heliopause 32
heliosphere 32
heliostat 130
helium flash 70
Henry Draper catalogue 97
Herbig-Haro object 99
Hermes 65
hertz 120
Hertzsprung-Russell diagram 71
High Energy Astronomy Observatory 189
high tide 37
high-velocity star 102
Hipparchus 191
Hirayama families 64
Hohmann orbit 23
hold 149
hold-down arm 149
horizon 10
horizontal parallax 12
hour angle 11
hour circle 11
Hubble classification 105
Hubble's constant 108
Hubble's law 108
Hubble Space Telescope 191
Hubble time 109
Hyades 89
hydrazine 144
hydrogen line 137
hyperbola 19
hyperbolic velocity 19
hyperboloid 128
hypergolic 145
hypothesis 193

IC 97
Icarus 65
igloo 173
ignition 145
image 123
image intensifier 133
immersion 42
impact 66
impact crater 34

inclination 19
inertial guidance 155
inertial measurement unit 155
Inertial Upper Stage 176
inferior conjunction 25
inferior planet 24
infinite 193
InfraRed Astronomy Satellite 190
infrared radiation 120
injection 151
injector 146
insertion 151
insolation 28
integration 193
Intelsat 180
interface 194
interference 120
interference fringes 120
interferometer 133
International Date Line 51
International Sun-Earth Explorer 191
International Ultraviolet Explorer 190
interplanetary 158
interstellar absorption 100
interstellar extinction 100
interstellar grain 100
interstellar medium 99
interstellar molecule 100
invariable plane 26
inverse-square law 114
ion 115
ionization 116
ionosphere 44
ion propulsion 143
iron meteorite 167
irregular galaxy 104
irregular variable 83
isotropic 110

jet 194
jet propulsion 141

jettison 151
Julian calendar 53
Julian date 53
Juno 65
Jupiter 58

K corona 32
Kelvin scale 118
Kepler's laws 17
Kirkwood gaps 64

Lacertid 107
Lagrangian points 22
Lalande 21185 94
Landsat 183
Large Magellanic Cloud 103
last contact 41
last quarter 34
late-type star 73
latitude 194
launch 148
launch azimuth 150
launch escape system 150
launch pad 148
launch vehicle 141
launch window 148
lava 194
layer 194
leap second 52
leap year 53
lens 123
lenticular galaxy 104
lepton era 111
libration 36
libration in latitude 36
libration in longitude 36
libration points 22
life support 163
lift-off 149
light 120
light curve 81
light time 91
light year 91
limb 26
limb darkening 28
limit 194
line of apsides 18

line of nodes 20
liquid fuel 144
Local Group 106
local standard of rest 90
local supercluster 106
longitude 194
longitude of perihelion 20
longitude of the ascending node 20
long module 172
long-period variable 83
lookback time 109
loop prominence *see* active prominence 32
low tide 37
luminosity 68
luminosity class 74
Luna 184
lunar 33
lunar calendar 52
lunar eclipse 38
lunar module 165
Lunar Orbiter 184
lunar roving vehicle 166
lunar year 54
lunation 34
lunisolar calendar 52
lunisolar precession 13
Lunokhod 185

Magellanic Cloud 103
Magellanic Stream 103
magnetic field 194
magnetic pole 46
magnetic star 84
magnetopause 44
magnetosphere 44
magnetotail 44
magnification 124
magnitude 68
main sequence 71
major axis 18
Maksutov telescope 130
manned manoeuvring unit 164
manoeuvre 156
mare 35

Mariner 185
Markarian galaxy 107
Mars 57
Mars probes 186
mascon 35
mass function 86
mass loss 74
mass-luminosity relation 71
mass ratio 142
mass transfer 87
matter era 110
maximum 194
max q 150
mean 194
mean anomaly 21
mean motion 21
mean solar time 49
mean Sun 49
mean time 49
mechanical 194
Mercury$_1$ 57
Mercury$_2$ 166
meridian 10
Messier catalogue 97
meteor 62
meteorite 66
meteoroid 62
meteor shower 62
meteor stream 62
Meteosat 182
Metonic cycle 52
microgravity 152
micrometeorite 66
microwaves 135
mid-course correction 158
mid-deck 171
military 194
Milky Way 101
Mills Cross 135
minimum 194
minimum-energy orbit 23
minor axis 19
minor planet 64
Mir 168
Mira 83
mirror 123
missile 141

220 · INDEX

missing mass 106
mission 153
mission control 153
mission specialist 173
mixture ratio 146
module 164
molecular line 137
molecule 116
momentum 194
mono-propellant 144
month 55
Moon 33
moon 25
Morgan-Keenan classification 73
motion 194
mounting 131
moving cluster 88
moving cluster method 91
multiple star 85
multi-stage rocket 141
mural quadrant 140

nadir 10
naked-eye star 69
naked singularity 80
navigate 195
navigation satellite 181
Navstar 181
neap tide 37
nebula 98
nebular hypothesis 27
negative ion 115
Neptune 59
neutral atom 116
neutral body position 163
neutral buoyancy tank 164
neutron 116
neutron star 76
new Moon 33
Newtonian reflector 128
N galaxy 106
NGC 97
nightglow 46
night sky light 46
Nimbus 182
noctilucent cloud 46

nocturnal 56
node 20
nominal 195
non-gravitational force 61
northern lights 45
nose-cone 142
nova 84
nozzle 147
nuclear-electric rocket 143
nuclear pulse rocket 143
nuclear reaction 70
nuclear rocket 143
nucleosynthesis 117
$nucleus^1$ 116
$nucleus^2$ 60
nutation 13

OB association 88
object glass 123
objective 123
objective prism 127
oblateness 26
obliquity of the ecliptic 9
observatory 137
observe 195
occultation 42
Olbers' paradox 109
Oort cloud 61
open cluster 88
open Universe 109
opposition 25
optical double 86
optics 123
orbit 16
orbital manoeuvring system 170
Orbiter 169
Orbiting Astronomical Observatory 188
Orbiting Solar Observatory 191
orientation 195
Orion arm 103
Orion nebula 99
oscillate 195
oscillating Universe 111
osculating elements 21

osculating orbit see osculating elements 21
oxidizer 144

Pallas 65
pallet 173
Palomar Sky Survey 97
parabola 19
parabolic velocity 19
paraboloid 128
parachute 161
parallax 91
parallel 195
parameter 195
parking orbit 23
parsec 91
partial eclipse 41
partial pressure 163
particle 195
payload 142
Payload Assist Module 176
payload bay 171
payload specialist 173
peculiar motion 90
peculiar velocity 90
$penumbra^1$ 41
$penumbra^2$ 29
penumbral eclipse 41
peri- 17
periapsis see apsis 18
periastron 17
perigee 17
perihelion 17
$period^1$ 20
$period^2$ 80
periodic comet 61
period-luminosity relation 82
Perseus arm 103
perturbation 21
phase 33
photocell 133
photoelectric cell 133
photographic magnitude 69
photographic zenith tube 56
photoionization 116

photometer 132
photomultiplier 133
photon 118
photon propulsion 143
photosphere 28
photovisual
   magnitude 69
physical double 86
Pioneer 186
Pioneer-Venus 186
pitch 156
pitchover 150
plage 31
Planck's law 122
plane 195
planet 24
planetarium 134
planetary nebula 100
planetary precession 13
planetesimal 27
Planet X 59
plasma 116
Platonic year 13
Pleiades 89
Pluto 59
pogo 142
polar axis 132
polar distance 11
polar flattening 26
polarized 122
polar orbit 23
pole 9
pole of the ecliptic 9
pole star 11
Population I 75
Population II 75
pore 29
portable life-support
   system 164
posigrade 157
position 8
position angle 15
positive ion 115
Poynting-Robertson
   effect 63
precession 13
precession of the
   equinoxes *see*
   precession 13
predict 195

preflare 162
pressure 195
primary^1 16
primary^2 86
prime focus 124
primeval fireball 110
prism 127
prismatic astrolabe 56
Procyon 94
prograde motion 14
prominence 31
propellant 143
proper motion 89
propulsion 141
proton 127
proton-proton chain 70
Proton rocket 177
protoplanet 27
protostar 69
Proxima Centauri 94
Ptolemaic system 139
pulsar 78
pulsate 195
pulsating variable
   star 81
pulse 195
pump 145

quadrant 140
quasar 107
quasi-stellar object 107
quasi-stellar radio
   source 107
quiescent
   prominence 31

radar astronomy 137
radial velocity 90
radian 15
radiant 63
radiate 119
radiation era 110
radiation pressure 61
radio astronomy 134
radio blackout 160
radio galaxy 107
radio interferometer 135
radio source 134
radio telescope 134
radio waves 120

radius 195
radius vector 17
Ranger 184
ratio 195
ray 35
ray crater 35
reaction control
   system 157
reaction propulsion 141
real time 154
recombination 116
reconnaissance
   satellite 183
record 195
recovery 161
recurrent nova 84
red dwarf 72
red giant 72
redshift 115
red spot 58
redundancy 142
re-entry 159
re-entry corridor 159
refer 195
reference 196
reflect 123
reflecting telescope 127
reflection nebula 99
reflector 127
refract 123
refracting
   telescope 127
refractor 127
region 196
regolith 35
regression of nodes 20
relativistic 113
relativity 113
remote manipulator
   system 171
remote sensing 183
rendezvous 157
research 196
residual 21
resolution 125
resolving power 125
restrict 196
retro-fire 160
retrograde motion 14
retro-rocket 160

222 · INDEX

revolution[1] 26
revolution[2] 153
Rigel 94
right ascension 8
rille 36
rima 36
Ring nebula 100
rings of Saturn 58
rise 11
Ritchey-Chrétien reflector 129
Roche's limit 59
Roche lobe 87
rocket 141
roll 156
roll reversal 161
rotation 26
RR Lyrae variable 82
RV Tauri stars 83
runaway star 77
runway 162

Sagittarius A 102
Sagittarius arm 103
Salyut 168
SAO catalogue 97
Saros 40
satellite[1] 24
satellite[2] 152
Saturn 58
Saturn IB 174
Saturn V 174
scattering 46
Schmidt-Cassegrain 130
Schmidt telescope 130
Schwarzschild radius 79
scintillation 47
Scout 175
scrub 149
season 54
secondary[1] 16
secondary[2] 86
secondary mirror 128
second contact 41
secular 196
secular parallax 91
seeing 47
selenology 36

semimajor axis 18
semi-regular variable 83
sensitive 196
separation 151
service module 165
service propulsion system 165
set 11
SETI 196
setting circle 132
Seyfert galaxy 106
shell star 84
shooting star 62
short module 172
short-period comet 61
shroud 142
shutdown 147
sidereal 48
sidereal day 48
sidereal month 55
sidereal period 25
sidereal time 48
sidereal year 54
siderite 67
siderolite 67
siderostat 130
signal 196
simulator 162
single-line spectroscopic binary 86
singularity 80
sinuous rille 36
Sirius 94
Sirius B 75
site 196
Skylab 168
Small Astronomy Satellite 188
small circle 10
Small Magellanic Cloud 103
solar 28
solar antapex 91
solar apex 90
solar calendar 52
solar cell 158
solar constant 28
solar day 48
solar eclipse 39

Solar Maximum Mission 191
solar motion 90
solar panel 158
solar parallax 24
solar power satellite 183
solar system 24
solar time 49
solar wind 32
solar year 54
solid fuel 145
Solid Rocket Booster 170
solstice 10
solstitial colure *see* colure 11
sounding rocket 141
source 119
southern lights 45
Southern Sky Survey 97
Soyuz 167
space 47
spacecraft 152
Spacelab 172
space probe 152
Space Shuttle 169
space sickness 163
space station 167
spacesuit 164
space-time 113
Space Tracking and Data Network 154
Space Transportation System 169
space velocity 90
special theory of relativity 113
specific impulse 145
speckle interferometry 133
spectral lines 121
spectral type 73
spectrogram 121
spectrograph 132
spectroheliograph 132
spectrometer 132
spectrophotometer 132
spectroscope 132
spectroscopic binary 86

spectroscopic parallax 91
spectroscopy 120
spectrum 119
spectrum binary 87
spectrum variable 84
speculum 123
sphere 196
spherical aberration 126
Spica 94
spicule 31
spin stabilization 157
spiral arm 104
spiral galaxy 104
spiral nebula 100
splashdown 161
sporadic 63
Spörer's law 30
spray prominence *see* active prominence 32
spring equinox 9
spring tide 37
Sputnik 178
stabilization 157
stage 141
staging 151
star 69
star atlas 96
star catalogue 96
star cluster 88
state vector 156
static firing 147
stationary point 14
statistical parallax 91
steady state theory 111
stellar association 88
stellar energy 69
stellar evolution 71
stellar population 75
stellar wind 75
step rocket 141
Stonehenge 139
stony-iron meteorite 67
stony meteorite 66
strap-on booster 142
Strömgren sphere 99
sub-dwarf 72
sub-giant 72
sub-orbital 151

summer solstice 10
Sun 28
sundial 56
sunrise 47
sunset 47
sunspot 29
sunspot cycle 29
sunspot maximum 29
sunspot minimum 29
Sun-synchronous orbit 23
supercluster 106
supergiant elliptical galaxy 104
supergiant star 72
supergranulation 31
superior conjunction 25
superior planet 24
supernova 76
supernova remnant 77
Surveyor 184
sustainer 142
swing arm 149
synchronous orbit 22
synchronous rotation 26
synchrotron radiation 118
synodic 55
synodic month 55
synodic period 25
syzygy 25

tactical air navigation 161
tail 60
tangential velocity 90
tank 145
Tarantula nebula 99
T association 88
tektite 67
telemetry 154
telescope 127
telluric 43
Telstar 179
temperature 118
terminal area energy management 161
terminator 34
terracing 34

terrae 35
terrestrial 43
terrestrial planet 24
test 196
theory 196
thermal protection system 169
third contact 41
third quarter 34
Thor 175
three-axis stabilization 158
throat 146
thrust 145
thruster 157
thrust vector control 147
tidal bulge 37
tidal distortion 37
tidal friction 37
tide 37
tilt 196
time dilation 113
time of perihelion passage 20
time zone 50
Tiros 182
Titan 175
topocentric coordinates 12
total eclipse 40
totality 40
touchdown 162
tracking 153
Tracking and Data Relay Satellite 181
trajectory 16
transfer ellipse *see* transfer orbit 23
transfer orbit 23
Transit 181
transit^1 11
transit^2 42
transit instrument 56
translation 157
transmit 196
transverse velocity 90
Trapezium 99
trigonometric parallax 91

triple alpha process 70
Trojans 65
tropic 10
tropical month 55
tropical year 53
tropic of Cancer 10
tropic of Capricorn 10
true anomaly 21
T Tauri stars 84
tuning-fork diagram 105
twilight 47
Tychonic system 139
Type I supernova 77
Type II supernova 77

UBV system 74
Uhuru 189
ullage 145
ultraviolet radiation 120
umbilical 148
umbilical tower 148
$umbra^1$ 41
$umbra^2$ 29
unit 196
Universal Time 49
Universe 108
uplink 155
Uranus 59
UV Ceti 83

vacuum 156
Van Allen belt 45
Vanguard 178
variable star 80
variation 36
vary 80
$Vega^1$ 94
$Vega^2$ 187
Veil nebula 100
Vela pulsar 78
velocity 196
velocity of light 120
Venera 186
Venus 57
vernal equinox 9
vernier rocket 142
very-long baseline
 interferometry 136
Vesta 65
Viking 186
virial theorem 106
Virgo cluster 106
visible 196
visual magnitude 68
volcano 196
volume 196
Voskhod 167
Vostok 167
Vostok rocket 177
Voyager 187

waning 34
waning crescent *see*
 crescent 34
wavelength 119
waxing 34
waxing crescent *see*
 crescent 34
weather satellite 182
weightlessness 152
white dwarf 75
Widmanstätten
 pattern 67
Wilson effect 30
winter solstice 10
Wolf 359 94
Wolf-Rayet stars 75
working fluid 144
world line 114
wrinkle ridge 35
W Virginis stars 82

X axis 156
X-rays 120
X-ray binary 87

yaw 156
Y axis 156
year 53

Z axis 156
Zeeman effect 122
zenith 10
zenithal hourly rate 63
zenith distance 11
zero *g* 152
zodiac 96
zodiacal light 46
Zond 185
zone of avoidance 103

The Solar System